A Comprehensive Guide to the Prostate

A Comprehensive Guide to the Prostate

Eastern and Western Approaches for Management of BPH

Edited by

Bilal Chughtai

Weill Cornell Medicine, New York, NY, United States

ELSEVIER

ACADEMIC PRESS

An imprint of Elsevier

Academic Press is an imprint of Elsevier
125 London Wall, London EC2Y 5AS, United Kingdom
525 B Street, Suite 1800, San Diego, CA 92101-4495, United States
50 Hampshire Street, 5th Floor, Cambridge, MA 02139, United States
The Boulevard, Langford Lane, Kidlington, Oxford OX5 1GB, United Kingdom

Notices

Knowledge and best practice in this field are constantly changing. As new research and experience broaden our understanding, changes in research methods, professional practices, or medical treatment may become necessary.

Practitioners and researchers must always rely on their own experience and knowledge in evaluating and using any information, methods, compounds, or experiments described herein. In using such information or methods they should be mindful of their own safety and the safety of others, including parties for whom they have a professional responsibility.

To the fullest extent of the law, neither the Publisher nor the authors, contributors, or editors, assume any liability for any injury and/or damage to persons or property as a matter of products liability, negligence or otherwise, or from any use or operation of any methods, products, instructions, or ideas contained in the material herein.

Library of Congress Cataloging-in-Publication Data
A catalog record for this book is available from the Library of Congress

British Library Cataloguing-in-Publication Data
A catalogue record for this book is available from the British Library

ISBN: 978-0-12-811464-3

Publisher: Mica Hayley
Acquisition Editor: Stacy Masucci
Editorial Project Manager: Sam Young
Production Project Manager: Mohana Natarajan
Designer: Matthew Limbert

Typeset by TNQ Books and Journals

Contents

List of Contributors

Khaled Ajib Centre Hospitalier de l'Université de Montréal (CHUM), Montréal, QC, Canada

Gina M. Badalato New York Presbyterian Hospital/Columbia University Medical Center, New York, NY, United States

Douglas C. Cheung University of Toronto, Toronto, ON, Canada

Bilal Chughtai Weill Cornell Medicine, New York, NY, United States

Caroline Chung Weill Cornell Medicine, New York, NY, United States

Doreen E. Chung Columbia University Medical Center/New York Presbyterian Hospital, New York, NY, United States

Niall F. Davis Beaumont Hospital, Co Dublin, Ireland

Erica L. Ditkoff Brigham & Womens Hospital, Boston, MA, United States

Dean Elterman University of Toronto, Toronto, ON, Canada

James C. Forde Beaumont Hospital, Co Dublin, Ireland

Pierre-Alain Hueber Centre Hospitalier de l'Université de Montréal (CHUM), Montréal, QC, Canada

Steven Kaplan Icahn School of Medicine, New York, NY, United States

Karthik R. Krishnan Weill Cornell Medicine, New York, NY, United States

Nathaniel D. Kwok Weill Cornell Medicine, New York, NY, United States

Matthew Lenardis University of Toronto, Toronto, ON, Canada

Michael J. Lipsky New York Presbyterian Hospital/Columbia University Medical Center, New York, NY, United States

Brady I. Magaoay Weill Cornell Medicine, New York, NY, United States

Malek Meskawi Centre Hospitalier de l'Université de Montréal (CHUM), Montréal, QC, Canada

Alexandra L. Millman University of Toronto, Toronto, ON, Canada

Carrie M. Mlynarczyk New York Presbyterian Hospital/Columbia University Medical Center, New York, NY, United States

H. Gregory Moore Weill Cornell Medicine, New York, NY, United States

Benjamin E. Muller New York Presbyterian Hospital/Columbia University Medical Center, New York, NY, United States

Cristina Negrean Centre Hospitalier de l'Université de Montréal (CHUM), Montréal, QC, Canada

Matthew P. Rutman New York Presbyterian Hospital/Columbia University Medical Center, New York, NY, United States

Zobia Shah Weill Cornell Medicine, New York, NY, United States

Alexander C. Small New York Presbyterian Hospital/Columbia University Medical Center, New York, NY, United States

Alexis Te Weill Cornell Medicine, New York, NY, United States

Dominique Thomas Weill Cornell Medicine, New York, NY, United States

Henry Tran Columbia University Medical Center/New York Presbyterian Hospital, New York, NY, United States; University of British Columbia, Vancouver, Canada

Stephanie Tsai Weill Cornell Medicine, New York, NY, United States

Roger Valdivieso Centre Hospitalier de l'Université de Montréal (CHUM), Montréal, QC, Canada

Sven J. Walderich Weill Cornell Medicine, New York, NY, United States

James P. Winebrake Weill Cornell Medicine, New York, NY, United States

Kevin C. Zorn Centre Hospitalier de l'Université de Montréal (CHUM), Montréal, QC, Canada

Introduction to a Comprehensive Guide to Your Prostate

1

Dominique Thomas, Bilal Chughtai
Weill Cornell Medicine, New York, NY, United States

Benign prostatic hyperplasia (BPH) and its sequela lower urinary tract symptoms are nondiscriminatory chronic conditions affecting approximately 50% of men as they reach their 50th decade of life.[1] As the aging male population increases, the incidence of BPH steadily increases linearly.[1] In the United States, each year trillions of dollars are spent on health expenditures to treat those with symptoms.[2] Furthermore, patients affected by BPH are known to be at higher risk of developing depression, having a decreased quality of life, falls, and increased mortality.[3] Thus, developing an armamentarium of treatment options, ranging from medical therapy to surgical procedures when the patient is symptomatic is a major medical concern.

A Comprehensive Guide to the Prostate: Eastern and Western Approaches for Management of BPH is a complete and multidisciplinary guide toward the diagnosis, management and treatment modalities associated with BPH. This guide will explore the different treatment options available for symptomatic patients from medical therapy and behavioral changes, naturopathic methods for patients with a preference for holistic approaches to their health and surgical methods.

First-line therapies include medications such as alpha-blockers, 5-alpha reductase inhibitors, or combination therapy, which has shown relief for certain patients.[4] However, patients may seek out other options such has naturopathic methods such as herbs to heal the prostate. Surgically, there have been a variety of techniques used to resect or ablate the enlarged prostatic tissue.

Each chapter will delve into the different management options available for BPH, which have been utilized and their demonstrated efficacy through the available research and clinical data. Furthermore, we will introduce alternative techniques used to treat BPH such as Eastern and Western medical practices. This compendium of writing is intended to shed light on complementary and alternative medicine as well as novel technologies employed in the treatment of BPH. The goal is to provide a comprehensive guide that will allow patients and caretakers, anyone treating BPH and researchers alike, to understand the BPH disease process and a comprehensive approach toward its treatment. Thus, this book is a unique and irreplaceable reference guide.

References

1. Chughtai B, Forde JC, Thomas DD, et al. Benign prostatic hyperplasia. *Nat Rev Dis Prim* 2016;**2**:16031.
2. Centers for Disease Control and Prevention: Health Expenditures. CDC.

A Comprehensive Guide to the Prostate. https://doi.org/10.1016/B978-0-12-811464-3.00001-5

3. Markland AD, Thompson IM, Ankerst DP, Higgins B, Kraus SR. Lack of disparity in lower urinary tract symptom severity between community-dwelling non-hispanic white, Mexican-American, and African-American men. *Urology* 2007;**69**(4):697–702.
4. Chughtai B, Thomas D, Kaplan S. Alpha-blockers, 5-alpha-reductase inhibitors, acetylcholine, beta3 agonists, and phosphodiesterase-5s in medical management of lower urinary tract symptoms/benign prostatic hyperplasia: how much do the different formulations actually matter in the classes? *Urol Clin N Am* 2016;**43**(3):351–6.

Behavioral Therapies and Self-Management for Benign Prostatic Hyperplasia

2

Dominique Thomas, Bilal Chughtai
Weill Cornell Medicine, New York, NY, United States

Introduction

For benign prostatic hyperplasia (BPH), one of the first-line therapies has traditionally been watchful waiting. Watchful watching is the conservative management of lower urinary tract symptoms (LUTS) through which the patient's symptoms are monitored for progression through infrequent visits to the urologist.[1] However, over the past several years, this practice has been on steady decline with other methods such as medical therapy and minimally invasive techniques being on the rise. Another area gleaning interest has been the use of behavioral modifications to help alleviate symptoms. The first-line treatment is now being described as consulting with a physician to reduce behaviors, which make a patient's BPH symptoms worse or increase behaviors, which help alleviate symptoms. Self-management can be divided into three sections: education and reassurance, behavioral interventions, and lifestyle modifications.[2] These methods can be very attractive to patients because it allows them to be active in their treatment while offering patients with low-risk symptom relief. The purpose behind this type of management is to reduce symptoms so medical interventions such as drugs are not required. Furthermore, these methods can be effective in reducing the number of resources used by the patient.[3] A stipulation to these methods is they can be very time-consuming for the patient and lead to burnout. The use of self-management varies from physician to physician leading to very few data existing on its benefits. The following chapter will navigate through the current suggestions. It will delve into the techniques the patient can employ when faced with uncomplicated LUTS by making modifications to their lifestyle.

Behavioral Changes

A practice commonly used by patients with mild-to-moderate symptoms for LUTS secondary to BPH is known as double voiding. This practice is used to ensure that the bladder is completely empty of urine.[4] The patient will go to the bathroom like usual to pass urine. After they are finished urinating, they may relax for a few moments and then attempt to urinate again. A variation of this practice involves leaving the bathroom and engaging in another activity. After a few minutes, the patient will go to the

A Comprehensive Guide to the Prostate. https://doi.org/10.1016/B978-0-12-811464-3.00002-7

bathroom again to urinate. Currently there have not been any formal studies conducted evaluating the efficacy of this method. Despite this, anecdotal evidence reveal this is a helpful method to those patients with irritative urinary symptoms.

Another method is known as bladder training or timed voiding. The purpose of this is to train the patient to use more normal patterns of urinating. Patients should not rely on only urinating when the urge is present. This method is effective in preventing the bladder from overfilling, while giving the patient more control over their bladder.[5] Timed voiding is often times classified as a type of bladder training. Many of the studies that have been conducted evaluating the efficacy of bladder have been done so on female patients with overactive bladder. However, a study by Zhengyong et al. investigated "efficacy of bladder training before removing the indwelling urinary catheter in patients with acute urinary retention (AUR) associated with benign prostatic hyperplasia (BPH)."[6] A total of 845 patients with a primary episode of AUR were randomized into the study. They were randomized into the following categories: "pharmacological treatment combined with bladder training or pharmacological treatment (tamsulosin 0.2 mg and finasteride 5 mg once daily) with free drainage of urinary catheter for 7 days, and a trial without catheter (TWOC) was performed."[6] Patients in the TWOC group had a 66.9% success rate after the 7 days of being catheterized. Furthermore, those in the intervention group along with TWOC had a 65.2% success rate and those who were in the control group were at 68.6%. Despite this, there were no statistically significant decreases in TWOC ($P > .05$).[6] This study demonstrated bladder training before the removal of a catheter did not increase the success of TWOC.

Brown et al. conducted a multidisciplinary panel that evaluated 94 action items that would be used during a self-management for men with LUTS secondary to BPH. Researchers used semistructured interviews and used practice surveys to evaluate the different behaviors. Items were presided over using the Research and Development Appropriateness Method. In total, 57 items were agreed on as follows: "patient assessment prior to starting a self-management programme (6), education and reassurance (4), fluid management (6), caffeine (4), alcohol (2), concurrent medication (2), types of toileting (2), bladder re-training (15), miscellaneous (1), and implementation of a self-management programme (15)."[7] A caveat to this study was these items were suggestions as to what options may be effective.

To provide evidence for these methods, Brown et al. conducted a pilot study of 25 men enrolled into a self-management program. Characteristics included a mean age of 64 years with uncomplicated LUTS. Furthermore, patients had an International Prostate Symptom Score (IPSS) at baseline to 19.6. The duration of the study was 3 months and patients were given "education (prostate and bladder), reassurance (expected future symptoms, prostate cancer risk), lifestyle modifications (fluid management by evening fluid restriction, caffeine and alcohol advice), and behavioral interventions (double-voiding, urethral milking, bladder retraining)" as methods to control their symptoms.[8] Patients were also asked to completed 3-day voiding diaries at baseline as well as at 1, 3, and 6 months. Those in this program were able to see a reduction in their episodes of nocturia, voids in 24 h, urgency. Furthermore, their IPSS saw a significant reduction to 8.5 at the 3-month follow-up.

Another study by Brown et al. evaluated the use of self-management in patients with LUTS as a first-line treatment. A total of 140 men were randomized into the trial to receive either self-management and standard care or standard care alone. Self-management was defined as "education, lifestyle advice, and training in problem solving and goal setting skills."[9] Treatment failure was measured across three time points: 3, 6, and 12 months. For patients in the self-management treatment arm at 3 months, treatment failure was 10% and standard care was 42%.[9]

At 6 months, treatment failure was 27% in self-management and 57% in standard care. At 12 months, treatment failure was 32% in self-management and 64% in standard care. Self-management was effective in reducing patients' urinary symptoms and the treatment failure.

Other Behavioral Modifications

As previously mentioned there have not been many research studies evaluating the effectiveness of behavioral modifications compared to other methods. Many of these suggestions should be used for men who currently do not have any complications associated with their LUTS secondary to BPH and may be in addition to pharmacologic interventions. Despite this, behavioral changes may be effective for all patients within the BPH cohort. A systematic review by de Jong et al. evaluated the clinical efficacy of urinating standing versus sitting in men with prostate enlargement in comparison to healthy males.[10] The following parameters were assessed maximum flow rate (Qmax), postvoid residual volume (PVR), and voiding time (TQ). A total of 11 articles were analyzed based on the inclusion criteria. Those with LUTS showed deceases in PVR (−24.96 mL); and TQ (−0.62 s) and an increase in Qmax (1.23 mL/s) while urinating in the sitting position.[10] However, these results were not statistically significant. Healthy men showed no difference in the parameters when urinating in the sitting or standing position Qmax (0.18 mL/s), TQ (0.49 s), and PVR (0.43 mL).[10]

Patients should be advised to also limit their amount of fluid consumption before going to bed and also when they are not going to be near a bathroom. Diuretics such alcohol and caffeine should also be avoided as these irritate the bladder causing urinary frequency.[11] Other diuretics come in the form of medication and can lead to worsening symptoms in men with LUTS and even lead to urinary retention.[12] The following list consists of other suggestions that can be used sparingly at the patient's discretion:

1. **Limit the use of decongestants or antihistamines:** These medications can lead to urinary retention as they obstruct the control of urine flow.
2. **Urinate when you first feel the urge:** If you hold your urine too long it may lead to large waist circumferences or a urinary incontinence episode.
3. **Eat a healthy diet and stay active:** Obesity is a comorbidity of BPH and staying fit can reduce the incidence of LUTS secondary to BPH.
4. **Keep yourself warm:** Colder temperatures are known to contribute to urinary urgency and frequency.

Lastly, pelvic floor muscle training may be effective. It can help with incontinence by preventing leaking. The purpose of these exercises is to strengthen the muscles, which support the bladder and help to close the sphincter so there is no leaking of urine.

As described by the University of Maryland Medical Center, the exercise should be performed as follows:

1. "Contract the muscle until the flow of urine is slowed or stopped. He attempts to hold each contraction for 20 s"[13]
2. "Release the contraction"[13]
3. "In general, patients should perform 5–15 contractions, three to five times daily"[13]

Conclusion

Self-management and behavioral modifications can be useful resources for men with LUTS secondary to BPH whose symptoms are uncomplicated. These programs need to be well designed depending on the patient's lifestyle. Patients with chronic conditions over time may have exacerbation of symptoms. Those treated with self-management will periodic reassignment of symptoms on a regular basis. Making behavioral modifications can be very helpful depending on the patient; however, there is a paucity of literature to support these claims, with the majority of information being anecdotal. Large-scale randomized controlled trials comparing the efficacy to controls and other treatment options are needed to understand the role of these methods and their feasibility.

References

1. McVary KT, Roehrborn CG, Avins AL, et al. Update on AUA guideline on the management of benign prostatic hyperplasia. *J Urol* May 2011;**185**(5):1793–803.
2. Brown CT, Emberton M. Could self-management challenge pharmacotherapy as a long-term treatment for uncomplicated lower urinary tract symptoms? *Curr Opin Urol* January 2004;**14**(1):7–12.
3. Newman S, Steed L, Mulligan K. Self-management interventions for chronic illness. *Lancet (Lond, Engl)* October 23–29, 2004;**364**(9444):1523–37.
4. *Benign prostatic hyperplasia (BPH)-Home treatment*. 2017.
5. *Timed voiding – national incontinence*. 2017.
6. Zhengyong Y, Changxiao H, Shibing Y, Caiwen W. Randomized controlled trial on the efficacy of bladder training before removing the indwelling urinary catheter in patients with acute urinary retention associated with benign prostatic hyperplasia. *Scand J Urol* August 2014;**48**(4):400–4.
7. Brown BN, Mani D, Nolfi AL, Moalli P. Impact of prolapse mesh properties on the host inflammatory response. *Int Urogynecol J Pelvic Floor Dysfunct* July 2014;(1):S226–7.
8. Brown CT, The Clinical Effectiveness Unit RCoSoEa, Van Der Meulen J, et al. Lifestyle and behavioural interventions for men on watchful waiting with uncomplicated lower urinary tract symptoms: a national multidisciplinary survey. *BJU Int* 2003;**92**(1):53–7.

9. Brown CT, Yap T, Cromwell DA, et al. Self management for men with lower urinary tract symptoms: randomised controlled trial. *BMJ Clin Res Ed* January 06, 2006;**334**(7583):25.
10. de Jong Y, Pinckaers JH, ten Brinck RM, Lycklama a Nijeholt AA, Dekkers OM. Urinating standing versus sitting: position is of influence in men with prostate enlargement. A systematic review and meta-analysis. *PLoS One* 2014;**9**(7):e101320.
11. *Medical treatment of benign prostatic hyperplasia – UpToDate*. 2017.
12. White WB, Moon T. Treatment of benign prostatic hyperplasia in hypertensive men. *J Clin Hypertens* 2017;**7**(4):212–7.
13. Simon H, Zieve D. *Benign prostatic hyperplasia*. 2012.

Medical Therapy for Benign Prostatic Hyperplasia

3

Niall F. Davis, James C. Forde
Beaumont Hospital, Co Dublin, Ireland

Abbreviations

5ARI 5-alpha reductase inhibitor
AUR Acute urinary retention
BPH Benign prostate hyperplasia
DHT Dihydrotestosterone
IPSS International Prostate Symptom Score
LUTS Lower urinary tract symptoms

Introduction

Traditionally, the standard medical treatment options for male patients with lower urinary tract symptoms (LUTS) due to benign prostatic hyperplasia (BPH) were α1-antagonists, 5-alpha reductase inhibitors (5ARIs), or a combination of both pharmacological agents.[1] Although these agents remain the mainstay for medically treating symptomatic BPH, several novel pharmacological agents have recently been introduced.[2] Selective α1-antagonists, phosphodiesterase-5 (PDE5) inhibitors, and antimuscarinic agents are now available for the management of symptomatic male LUTS due to BPH. Introducing these agents into the pharmacological armamentarium is important as conventional therapies are associated with significant adverse effects and variable efficacy, which leads to poor compliance among male patients. Furthermore, the selection of newly available pharmacological agents for LUTS due to BPH can facilitate an individualized patient approach for managing symptomatic BPH.

The indications for surgery for symptomatic BPH have changed during the last 30 years due to the increasing use of medical therapy. Alpha-adrenergic antagonists and 5ARIs are now frequently used as first-line management in isolation or in combination to treat LUTS secondary to BPH when lifestyle modifications fail to improve the patient's symptoms. Therefore, urologists should have a thorough understanding on pharmacodynamics, clinical indications, and adverse effects when medical therapies are prescribed for treatment.[3] This chapter provides an overview on medical therapy for symptomatic LUTS due to BPH. We place particular emphasis on novel pharmacological agents in the setting of LUTS due to BPH.

A Comprehensive Guide to the Prostate. https://doi.org/10.1016/B978-0-12-811464-3.00003-9

Pathogenesis of Benign Prostatic Hyperplasia

Although symptomatic BPH was initially described in 1649; its pathophysiology remains poorly understood.[4,5] The prostate gland is the only male organ that continues to grow with age.[5] This unique feature has led some researchers to believe that BPH occurs due to reactivation of a dormant embryonic growth process within the gland's stromal tissue.[6,7] Interactions between the stroma and epithelium are controlled by the androgen receptor and dihydrotestosterone (DHT).[8] Both are heavily implicated in promoting prostate tissue expansion that forms characteristic enlarging benign nodules.[9] Ninety percent of serum testosterone is converted into the more potent hormone DHT in the prostatic tissues by the enzyme 5-alpha reductase-2 (5AR2). The significance of the relationship between testosterone and DHT is evident in medically castrated BPH patients. An improvement in LUTS is directly attributable to a decrease in prostate size in patients who are treated with 5ARIs for at least 3 months. Notably, this pathway is not the only regulator of prostate growth as a significant proportion of patients do not respond to 5ARIs.[10]

Prostatic tissue expansion results from paracrine cellular alterations that include proliferation, quiescence, and apoptosis. Growth factors such as fibroblast growth factor, epithelial growth factor, insulin-like growth factor, keratinocyte growth factor, hepatocyte growth factor, and vascular endothelial growth factor are all upregulated by DHT in BPH.[11,12] Other mechanisms linked to progression of BPH with age are increased levels of estrogen, loss of cell cycle regulators, and increased oxidative stress.[13–15] In addition, persistence of chronic inflammation has been linked to symptomatic BPH due to increased expression of cytokines such as interleukin-8 and hypoxia-inducible factor 1-α.[16–19] Lifestyle factors and comorbid diseases such as obesity, diabetes mellitus, and hypercholesterolemia have also recently been implicated in the pathogenesis of BPH through upregulation of the aforementioned growth factors.[20,21]

Alpha-Blocker Monotherapy

Adrenoceptors located within the stroma and smooth muscle of the prostate gland and bladder neck are responsible for contractility.[22] Alpha 1_A is the predominate receptor in prostate stromal smooth muscle, and nonselective α1-antagonists act on the dynamic component of BPH by inhibiting alpha-1_A, alpha-1_B, and alpha-1_D receptors.[22,23] Alpha adrenoceptors are composed of seven transmembrane domains and are coupled with G proteins. When stimulated, phospholipase C is activated with a resultant increase in intracellular Ca^{2+}, which stimulates contraction in smooth muscular cells.[22,24] This adrenoceptor targeting mechanism leads to an improvement in urinary flow. However, alpha-blocker therapy does not decrease prostate volume or the risk of disease progression or acute urinary retention (AUR).[25,26] The first generation alpha-blockers were nonselective to prostate adrenoceptors and were associated with cardiovascular and gastrointestinal adverse effects. Second generation alpha-blockers (e.g., doxazosin and terazosin) require dose titration due to their association with hypotension and cardiovascular risk factors. Third generation alpha-blockers are better tolerated as they are more selective for prostate adrenoceptors and do not require dose titration.

Currently available alpha 1-blockers are terazosin, doxazosin, alfuzosin, tamsulosin, silodosin, and naftopidil. Their doses and side effects are summarized in Table 3.1. All are effective for treating patients with mild, moderate, or severe LUTS due to BPH (i.e., volume <40 g).[27] Their side effects differ; however, all appear equivalent in terms of clinical effectiveness.[28] Randomized controlled trials (RCTs) have demonstrated that alpha 1-blockers can decrease International Prostate Symptom Score (IPSS) by 35%–40%, increase bladder capacity, decrease detrusor overactivity, and increase maximum urinary flow rate (Qmax) by 20%–25%.[29–31] Tamsulosin and silodosin are selective adrenoceptor blockers and are therefore prescribed

Table 3.1 Summary of doses and side effects for alpha-blockers used as monotherapy to treat symptomatic lower urinary tract symptoms due to benign prostatic hyperplasia

Name	Dose (mg/day)	Uroselective adrenoceptor blockade	Adverse effects
Doxazosin	1–8	No	Requires titration Dizziness Tachycardia Hypotension Syncope Headache Rhinitis
Terazosin	1–20	No	Requires titration Dizziness Tachycardia Hypotension Syncope Headache Rhinitis
Alfuzosin	10	Yes	Dizziness Headache Asthenia
Tamsulosin	0.4	Yes	Ejaculatory dysfunction Rhinitis
Silodosin	4–8	Yes	Highest rate of ejaculatory dysfunction
Naftopidil[a]	25–75	Yes	Required dose titration Ejaculatory dysfunction

[a]Only clinically available in Japan.

more frequently.[32] Tamsulosin is mildly selective for the alpha-1_A receptor compared to the alpha-1_B adrenoceptor; however, it is equally selective for alpha-1_A receptor and alpha-1_D receptors.[32] In comparison, silodosin is superselective for the alpha-1_A receptor compared to the alpha-1_B adrenoreceptor (162:1).[33,34] Both agents are associated with an improvement in LUTS after 4–5 days of treatment, which can be maintained for ≥12 months.[35] The most frequent side effects associated with alpha-blockers therapy are dizziness and postural hypotension due to their systemic profile. Typically, alpha 1-blockers do not affect libido but are associated with retrograde ejaculation.[32]

Five-α-Reductase Inhibitor Monotherapy

Steroid 5α-reductase inhibitors prevent the conversion of testosterone to DHT by inhibiting the enzyme 5AR2 (Fig. 3.1). This mechanism causes apoptosis of epithelial cells, which in turn leads to a decrease in prostate size.[36] Two 5ARIs are currently available for treating symptomatic LUTS due to BPH: dutasteride and finasteride. Their dose, mechanism of action and adverse effects are summarized in Table 3.2. The benefits of these agents become evident after 4–6 weeks with maximal efficacy usually occurring after 3–6 months compared to treatment with a placebo. Improvements in LUTS with 5ARIs are durable with a reduction in IPSS of 15%–30%, decreased prostate volume by ≥30% and increased Qmax on uroflowmetry by 2.0 mL/s after ≥2 years of treatment.[37–39] 5ARIs also play a role in treating hematuria due to BPH with 75% of patients experiencing no further bleeding after a mean follow-up of 3 years as demonstrated in a study by Kearney et al. on 53 male patients.[40] The reduction in IPSS with finasteride is dependent on the initial prostate size and symptom improvement becomes less apparent in prostate glands that are ≤40 g.[41,42] In comparison, dutasteride decreases IPSS, prostate volume, and the risk of AUR irrespective of prostate size.[43,44] Finasteride and dutasteride are associated with adverse effects such as decreased libido, decreased ejaculate volume, erectile dysfunction (ED) in 5%, breast symptoms in 1% and a rash.[45]

The Proscar Long-Term Efficacy and Safety Study evaluated the efficacy of finasteride in over 3000 patients.[46] Inclusion criteria were men with moderate to severe LUTS on IPSS, reduced flow rate and prostate-specific antigen <10 ng/μL. After a follow-up period of 4 years, prostate volume reduced by 18%, symptom score improved by 1.6 points and urinary flow rate improved by approximately 2 mL/s in the finasteride arm.[46] Furthermore, the relative risk of AUR and need for surgery decreased by approximately 55%. Adverse effects included reduced libido (6.4%), impotence (8.1%), reduced ejaculate volume (3.7%), rash (<1%), and breast tenderness (<1%).[46]

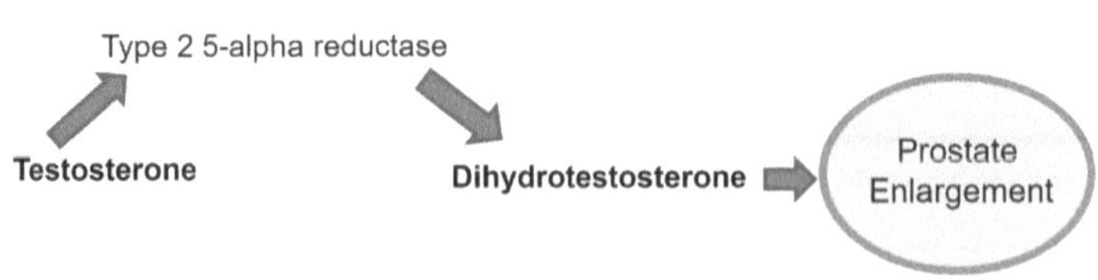

Figure 3.1 Conversion of testosterone to dihydrotestosterone by 5-alpha reductase type 2.

Table 3.2 Summary of doses and side effects for 5-alpha reductase inhibitors used as monotherapy to treat symptomatic lower urinary tract symptoms due to benign prostatic hyperplasia

Name	Dose (mg/day)	Mechanism of action	Adverse effects
Finasteride	5	Inhibits 5-α-reductase type 1	Decreased libido Erectile dysfunction Ejaculatory dysfunction Gynecomastia Breast tenderness Slow onset
Dutasteride	0.5	Inhibits 5-α-reductase type 1 and type 2	Decreased libido Erectile dysfunction Ejaculatory dysfunction Gynecomastia Breast tenderness Slow onset

Alpha-Blocker Monotherapy Combined With 5-α-Reductase Inhibitor

The role of combining α-blocker therapy with 5ARIs is established and recommended in all major guidelines for moderately/severely symptomatic men with LUTS due to BPH.[47,48] The main advantage of combination therapy is that short- and long-term management is instigated in patients. Combination therapy also plays an important role in treating patients with an increased risk of disease progression.[49] The established role for combination therapy is based on large-scale RCTs such as the Medical Therapy of Prostate Symptoms (MTOPS) and CombAT trials.[49,50]

The MTOPS study evaluated and compared the efficacy of a placebo, finasteride 5 mg, doxazosin 8 mg and a combination of doxazosin and finasteride in over 3000 patients with a 4.5-year follow-up period.[49] The primary end point was time to clinical progression, which was defined by at least 1 of the following criteria:

- IPSS ≥4-point increase (on two occasions within 4 weeks)
- AUR
- Recurrent urinary tract infections
- Renal insufficiency (≥50% rise in baseline serum creatinine and ≥1.5 mg/dL [creatinine 133 μg/L])
- Incontinence

After the follow-up period, only 17% of patients in placebo group progressed. 5% of patients progressed with combination therapy compared to 10% for doxazosin and finasteride alone. Apart from demonstrating the effectiveness of combination therapy, this trial also demonstrated important data on the slow progressive natural history of BPH in its placebo arm.[49]

The ComBAT study analyzed the combination of dutasteride and tamsulosin versus either drug alone in 4844 men with LUTS. Inclusion criteria for this trial were men >50 years of age with an IPSS ≥12, prostate volume ≥30 mL, and maximum flow rate (Qmax) between 5 and 15 mL/s.[50] The primary end point was development of AUR or necessity for surgical intervention. After 4 years of follow-up, those on combination therapy had better associated symptoms when compared to either drug alone. Dutasteride was associated with an improvement in flow rate (+18%) and a decrease in prostate volume (−22.8%) but had no significant additional effect with combination therapy (−22.8% vs. −23.4%, respectively).[50] A criticism of this study is the absence of a placebo arm so the "placebo effect" cannot be quantified with any of the pharmacological agents assessed. Metaanalyses of alpha-blocker single therapy and combination therapy of alpha-blockers and 5ARIs demonstrated a nonsignificant improvement in IPSS of −0.49 (95% CI, −1.01 to 0.02) with combination therapy compared to monotherapy.[37,38,51] However, there was a statistically significant improvement in Qmax of 0.88 mL/s (95% CI, 0.40 to 1.35) for combination therapy compared to monotherapy.[37,38,51] These data further support the use of long-term combination therapy with α-blocker therapy and 5ARIs in patients who are considered high risk for disease progression.

Phosphodiesterase Type 5 Inhibitor

PDE5 inhibitors were initially introduced for the treatment of ED; however, studies have also demonstrated their role in treating LUTS due to BPH.[1] PDE5 inhibitors increase cyclic guanosine monophosphate, which results in a reduction of the smooth muscle tone of the prostate through a nitric oxide (NO)-mediated reaction.[52] This reaction also causes decreased smooth muscle tone within the prostate and urethra, which may lead to an improvement in LUTS.[53] Currently, five PDE5 inhibitors are available: sildenafil, tadalafil, vardenafil, udenafil, and mirodenafil. Sildenafil was the first developed PDE5 inhibitor shown to improve LUTS due to BPH as demonstrated by objective decreases IPSS and improved quality of life (QoL) scores.[54,55] Studies evaluating the efficacy of other PDE5 inhibitors have demonstrated improvements in IPSS, maximum urine flow rate (Qmax), and decreased residual volume after micturition.[56–58]

Tadalafil is the most commonly prescribed PDE5 inhibitor for LUTS.[59] It is the only agent that has been approved by European and American guidelines for this purpose and is based on level 1b evidence from RCTs.[56,59,60] One randomized placebo-controlled trial demonstrated that tadalafil was equally effective with α-blocker monotherapy for treating LUTS due to BPH. Roehrborn et al. analyzed four large placebo controlled trials of tadalafil in 1500 male patients with LUTS.[50] After a follow-up period of 3 months, the median Qmax increased by 0.9 mL/s in the placebo arm and by 1.2 mL/s in the arm treated with tadalafil in patients with baseline voided volume of 150–250 mL. Overall, there was a small, but statistically significant improvement in Qmax with tadalafil compared to placebo. The significance of improved Qmax increased in patients with greater baseline voiding volumes (i.e., 350–450 mL).

Adverse effects associated with tadalafil include headache, facial flushing, back pain, and dyspepsia. Importantly, tadalafil is contraindicated in patients on nitrate therapy due to the associated risk of hypotension.[50]

Combination of Phosphodiesterase-5 Inhibitors With 5-α-Reductase Inhibitor

The combination of tadalafil combined with a 5ARI is appealing for several reasons. 5ARIs have a proven positive effect on the natural history of BPH progression; however, they are disadvantaged by their delayed pharmacokinetics and negative effects on sexual function[2] (Table 3.2). Tadalafil counteracts these disadvantages due to its positive effect on erectile function and rapid clinical onset. One randomized, double-blinded placebo-controlled trial[26] compared the effectiveness of tadalafil (5 mg) plus finasteride (5 mg) versus placebo versus finasteride (5 mg) in 695 patients.[61] Patients were randomized into two treatment groups with some men on both groups having a history of preexisting ED. International Index of Erectile Function scores were utilized to identify patients with ED. Findings demonstrated that tadalafil combined with finasteride was associated with an improvement in LUTS due to BPH regardless of preexisting ED.[61]

Another double-blinded randomized controlled trial also compared the effectiveness of tadalafil (5 mg) plus finasteride (5 mg) versus placebo versus finasteride (5 mg) in men with an IPSS ≥13 and a prostate volume ≥30 g.[62] The mean change in IPSS after coadministration of tadalafil plus finasteride was 5.5 after 26 weeks compared to 4.5 in the group that was coadministered tadalafil with a placebo. Furthermore, erectile function improved in the tadalafil plus finasteride group and the authors concluded by suggesting a role for this type of combination therapy in men with LUTS that are at an increased risk of disease progression who also suffer from ED.[62] Combination therapy with a PDE5 inhibitor and α-blocker is less supported in the literature due to conflicting results in trials with lower levels of evidence.[63]

Antimuscarinic Agent Combined With α-Blocker

The two predominant subtypes of muscarinic receptor expressed in the lower urinary tract are M2 and M3 receptors. The proportion of M2 receptors in detrusor membranes is 71% and M3 receptors are present in 22%.[64] M3 receptors are accountable are mainly for detrusor contraction M2 receptors are predominant in the urothelium and are associated with pathological changes in the bladder.[65] Antimuscarinic agents are widely used for treating patients with overactive bladder (OAB) symptoms by reducing detrusor contractility. Storage symptoms with urodynamically proven OAB can coexist in approximately 50% of patients with LUTS due to BPH.[66] These symptoms persist in approximately 35%–40% after treatment of bladder outlet obstruction. Combination treatment with α-blocker therapy and antimuscarinic agents have demonstrated greater efficacy for treating urgency and urge incontinence with improved QoL.[67] A summary of dosage, mechanism of action, and adverse effects for antimuscarinic agents and beta-3 (β_3) agonists used to treat OAB symptoms in patients with BPH is detailed in Table 3.3.

The TIMES study included 879 men with symptoms of BPH and OAB.[68] Patients were randomized to receive either tolterodine 4 mg with tamsulosin, tolterodine 4 mg alone, tamsulosin alone, or a placebo. The primary end point of the trial was patient perception of treatment benefit. The combination arm of the trial reported a significant decrease in urgency incontinence episodes (−0.88 vs. −0.31, P=.005), frequency (−2.54 vs. −1.41, P<.001), micturitions per 24 h, micturitions per night, and an improvement in QoL after 12 weeks compared to the placebo arm.[68] The rate of AUR was 1.8% in the placebo group, 0% in the α-blocker monotherapy group, 0.9% for the combination group, and 1.9% for the tolterodine monotherapy group. Dry mouth was the most significant side effect from the antimuscarinic agent and it occurred in 21% of patients in the combination group compared to 7% in each monotherapy group and 2% in the placebo arm.[68]

Table 3.3 Summary of doses and side effects for antimuscarinic agents/β_3 agonists used as combination therapy to treat symptomatic lower urinary tract symptoms with overactive bladder due to benign prostatic hyperplasia

Name	Dose (mg/day)	Mechanism of action	Notes
Darifenacin	7.5–15	Antimuscarinic	Administer with caution in patients with hepatic impairment
Fesoterodine	4–8	Antimuscarinic	Contraindicated with severe hepatic impairment
Oxybutynin (immediate release)	5–20	Antimuscarinic	Requires titration relative to side effects
Solifenacin	5–10	Antimuscarinic	Administer with caution in patients with hepatic impairment
Tolterodine	1–4	Antimuscarinic	Administer lower doses in patients with renal or hepatic dysfunction
Trospium	20–60	Antimuscarinic	Use with caution in patients with renal impairment
Mirabegron	25–50	β_3-agonist	Use with caution in patients with hepatic impairment

Future Perspectives and Conclusions

Research in the field of BPH therapy is continuously evolving. This paradigm shift is due to the complexity of the pathophysiology of BPH as the condition is not purely confined to the prostate gland. Alpha-blockers used as monotherapy act on the dynamic component of BPH by inhibiting alpha 1_A, alpha 1_B, and alpha-1_D receptors. Their efficacy is greatest when treating symptomatic BPH in prostate glands that are relatively small (i.e., <40 g). Recently developed alpha-blockers are more selective for prostate adrenoceptors and are associated with improved compliance among patients due to their lower side effect profile. In contrast to alpha-blockers, 5ARIs should be offered to men with moderate to severe LUTS and enlarged prostate volume (i.e., ≥40 g) detected by digital rectal examination and/or ultrasonography to reduce the risk of disease progression, AUR, and surgery. Combining α-blocker therapy with 5ARIs is recommended for male patients with moderate to severe LUTS due to BPH. This form of combination medical therapy is advantageous in BPH due to immediate and long-term symptomatic improvement with an associated decreased risk for disease progression. Tadalafil prescribed as monotherapy or in combination with a 5-ARI improves BPH symptoms in patients with and without a history of preexisting ED. Finally, antimuscarinic agents or beta-3 agonists combined with an alpha-blocker are most effective for treating OAB symptoms that are persistent in approximately 50% of patients after treatments of LUTS with alpha-blocker therapy.

Medical therapy remains the cornerstone for treating patients with symptomatic LUTS due to BPH. A variety of pharmacological agents are available as monotherapy or in combination that can be tailored to meet a patient's individual needs. In the future, a patient specific approach will enable individualized targeted therapy with maximal symptomatic improvement and minimal adverse effects.

References

1. Albisinni S, Biaou I, Marcelis Q, Aoun F, De Nunzio C, Roumeguère T. New medical treatments for lower urinary tract symptoms due to benign prostatic hyperplasia and future perspectives. *BMC Urol* 2016;**16**(1):58. [Internet]. Available from: http://bmcurol.biomedcentral.com/articles/10.1186/s12894-016-0176-0.
2. Schauer I, Madersbacher S. Medical treatment of lower urinary tract symptoms/benign prostatic hyperplasia. *Curr Opin Urol* 2015;**25**(1):6–11. [Internet]. Available from: http://content.wkhealth.com/linkback/openurl?sid=WKPTLP:landingpage&an=00042307-201501000-00003.
3. Chughtai B, Forde JC, Thomas DDM, Laor L, Hossack T, Woo HH, et al. Benign prostatic hyperplasia. *Nat Rev Dis Prim* 2016;**2**:16031. [Internet]. Available from: http://www.nature.com/articles/nrdp201631.
4. Manuscript A, Nanostructures SPC. NIH public access. *Nano* 2008;**6**(9):2166–71.
5. Loeb S, Kettermann A, Carter HB, Ferrucci L, Metter EJ, Walsh PC. Prostate volume changes over time: results from the Baltimore Longitudinal Study of Aging. *J Urol* 2009;**182**(4 Suppl.):1458–62.

6. Bushman W. Etiology, epidemiology, and natural history of benign prostatic hyperplasia. *Urol Clin North Am* 2009;**36**(4):403–15. v. [Internet]. Available from: http://www.ncbi.nlm.nih.gov/pubmed/19942041.
7. McNeal JE. Origin and evolution of benign prostatic enlargement. *Invest Urol* 1978;**15**(4):340–5. [Internet]. Available from: http://www.ncbi.nlm.nih.gov/pubmed/75197.
8. Roehrborn CG. Benign prostatic hyperplasia: an overview. *Rev Urol* 2005;**7**(Suppl. 9):S3–14. [Internet]. Available from: http://www.ncbi.nlm.nih.gov/pmc/articles/PMC1477638/.
9. Cunha GR, Fujii H, Neubauer BL, Shannon JM, Sawyer L, Reese BA. Epithelial-mesenchymal interactions in prostatic development. I. Morphological observations of prostatic induction by urogenital sinus mesenchyme in epithelium of the adult rodent urinary bladder. *J Cell Biol* 1983;**96**(6):1662–70.
10. McConnell JD, Bruskewitz R, Walsh P, Andriole G, Lieber M, Holtgrewe HL, et al. The effect of finasteride on the risk of acute urinary retention and the need for surgical treatment among men with benign prostatic hyperplasia. *J Urol* 1998:272.
11. Li W, Wu C-L, Febbo PG, Olumi AF. Stromally expressed c-Jun regulates proliferation of prostate epithelial cells. *Am J Pathol* 2007;**171**(4):1189–98.
12. Luo J, Dunn T, Ewing C, Sauvageot J, Chen Y, Trent J, et al. Gene expression signature of benign prostatic hyperplasia revealed by cDNA microarray analysis. *Prostate* 2002;**51**(3):189–200. [Internet]. Available from: http://doi.wiley.com/10.1002/pros.10087.
13. Chang YL, Lin ATL, Chen KK, Chang YH, Wu HHH, Kuo JY, et al. Correlation between serum prostate specific antigen and prostate volume in Taiwanese men with biopsy proven benign prostatic hyperplasia. *J Urol* 2006;**176**(1):196–9.
14. Nicholson TM, Ricke WA. Androgens and estrogens in benign prostatic hyperplasia: past, present and future. *Differentiation* 2011:184–99.
15. Ahmad M, Suhail N, Mansoor T, Banu N, Ahmad S. Evaluation of oxidative stress and DNA damage in benign prostatic hyperplasia patients and comparison with controls. *Indian J Clin Biochem* 2012;**27**(4):385–8. [Internet]. Available from: http://www.pubmedcentral.nih.gov/articlerender.fcgi?artid=3477465&tool=pmcentrez&rendertype=abstract.
16. McLaren ID, Jerde TJ, Bushman W. Role of interleukins, IGF and stem cells in BPH. *Differentiation* 2011:237–43.
17. Schauer IG, Rowley DR. The functional role of reactive stroma in benign prostatic hyperplasia. *Differentiation* 2011;**82**(4–5):200–10. [Internet]. Available from: http://www.sciencedirect.com/science/article/pii/S0301468111000727.
18. Chughtai B, Lee R, Te A, Kaplan S. Role of inflammation in benign prostatic hyperplasia. *Rev Urol* 2011;**13**(3):147–50. [Internet]. Available from: http://www.ncbi.nlm.nih.gov/pubmed/22110398%5Cnhttp://www.pubmedcentral.nih.gov/articlerender.fcgi?artid=PMC3221555.
19. Bostanci Y, Kazzazi A, Momtahen S, Laze J, Djavan B. Correlation between benign prostatic hyperplasia and inflammation. *Curr Opin Urol* 2013;**23**(1):5–10. [Internet]. Available from: http://content.wkhealth.com/linkback/openurl?sid=WKPTLP:landingpage&an=00042307-201301000-00004.
20. Hammarsten J, Hogstedt B. Calculated fast-growing benign prostatic hyperplasia–a risk factor for developing clinical prostate cancer. *Scand J Urol Nephrol* 2002;**36**(5):330–8. [Internet]. Available from: http://www.ncbi.nlm.nih.gov/entrez/query.fcgi?cmd=Retrieve&db=PubMed&dopt=Citation&list_uids=12487736.
21. Parsons JK, Sarma AV, McVary K, Wei JT. Obesity and benign prostatic hyperplasia: clinical connections, emerging etiological paradigms and future directions. *J Urol* 2013;**189**(1 Suppl.).
22. Michel MC, Vrydag W. Alpha1-, alpha2- and beta-adrenoceptors in the urinary bladder, urethra and prostate. *Br J Pharmacol* 2006;**147**(Suppl.):S88–119. [Internet]. Available from: http://www.pubmedcentral.nih.gov/articlerender.fcgi?artid=1751487&tool=pmcentrez&rendertype=abstract.

23. Oelke M, Bachmann A, Descazeaud A, Emberton M, Gravas S, Michel MC, et al. EAU guidelines on the treatment and follow-up of non-neurogenic male lower urinary tract symptoms including benign prostatic obstruction. *Eur Urol* 2013:118–40.
24. Kojima Y, Kubota Y, Sasaki S, Hayashi Y, Kohri K. Translational pharmacology in aging men with benign prostatic hyperplasia: molecular and clinical approaches to alpha1-adrenoceptors. *Curr Aging Sci* 2009;**2**(3):223–39.
25. McVary KT. Alfuzosin for symptomatic benign prostatic hyperplasia: long-term experience. *J Urol* 2006:35–42.
26. Vallancien G, Emberton M, Alcaraz A, Matzkin H, Van Moorselaar RJA, Hartung R, et al. Alfuzosin 10 mg once daily for treating benign prostatic hyperplasia: a 3-year experience in real-life practice. *BJU Int* 2008;**101**(7):847–52.
27. Djavan B. 1-Adenoceptor antagonists for the treatment of lower urinary tract symptoms suggestive of benign prostatic hyperplasia (LUTS/BPH): state of the art. *Eur Urol Suppl* 2004:23–30.
28. Lepor H. Alpha blockers for the treatment of benign prostatic hyperplasia. *Rev Urol* 2007;**9**(4):181–90. [Internet]. Available from: http://www.ncbi.nlm.nih.gov/pubmed/18231614%0Ahttp://www.pubmedcentral.nih.gov/articlerender.fcgi?artid=PMC2213889.
29. Lukacs B, Blondin P, MacCarthy C, Du Boys B, Grippon P, Lassale C. Safety profile of 3 months' therapy with alfuzosin in 13,389 patients suffering from benign prostatic hypertrophy. *Eur Urol* 1996;**29**:29–35.
30. Roehrborn CG, Rosen RC. Medical therapy options for aging men with benign prostatic hyperplasia: focus on alfuzosin 10 mg once daily. *Clin Interv Aging* 2008:511–24.
31. Nickel JC, Sander S, Moon TD. A meta-analysis of the vascular-related safety profile and efficacy of α-adrenergic blockers for symptoms related to benign prostatic hyperplasia. *Int J Clin Pract* 2008;**62**(10):1547–59.
32. van Dijk MM, de la Rosette JMCH, Michel MC. Effects of alpha(1)-adrenoceptor antagonists on male sexual function. *Drugs* 2006;**66**(3):287–301.
33. Yamanishi T, Mizuno T, Tatsumiya K, Watanabe M, Kamai T, Yoshida K. Urodynamic effects of silodosin, a new alpha 1A-adrenoceptor selective antagonist, for the treatment of benign prostatic hyperplasia. *Neurourol Urodyn* 2010;**29**(4):558–62. [Internet]. Available from: http://www.ncbi.nlm.nih.gov/entrez/query.fcgi?cmd=Retrieve&db=PubMed&dopt=Citation&list_uids=19693954.
34. Matsukawa Y, Gotoh M, Komatsu T, Funahashi Y, Sassa N, Hattori R. Efficacy of silodosin for relieving benign prostatic obstruction: prospective pressure flow study. *J Urol* 2009;**182**(6):2831–5.
35. Kawabe K, Yoshida M, Homma Y. Silodosin, a new alpha1A-adrenoceptor-selective antagonist for treating benign prostatic hyperplasia: results of a phase III randomized, placebo-controlled, double-blind study in Japanese men. *BJU Int* 2006;**98**(5):1019–24.
36. Rittmaster RS, Norman RW, Thomas LN, Rowden G. Evidence for atrophy and apoptosis in the prostates of men given finasteride. *J Clin Endocrinol Metab* 1996;**81**(2):814–9.
37. Roehrborn CG, Siami P, Barkin J, Damião R, Major-Walker K, Morrill B, et al. The effects of dutasteride, tamsulosin and combination therapy on lower urinary tract symptoms in men with benign prostatic hyperplasia and prostatic enlargement: 2-year results from the CombAT study. *J Urol* 2008;**179**(2):616–21.
38. Roehrborn CG, Siami P, Barkin J, Damião R, Major-Walker K, Nandy I, et al. The effects of combination therapy with dutasteride and tamsulosin on clinical outcomes in men with symptomatic benign prostatic hyperplasia: 4-year results from the CombAT study. *Eur Urol* 2010;**57**(1):123–31.
39. Keating GM. Dutasteride/tamsulosin: in benign prostatic hyperplasia. *Drugs Aging* 2012;**29**(5):405–19. [Internet]. Available from: http://www.ncbi.nlm.nih.gov/pubmed/22550968.

40. Kearney MC, Bingham JB, Bergland R, Meade-D'Alisera P, Puchner PJ. Clinical predictors in the use of finasteride for control of gross hematuria due to benign prostatic hyperplasia. *J Urol* 2002;**167**(6):2489–91. [Internet]. Available from: file:///Users/leezhao/Documents/Papers2/2002/Kearney/J Urol 2002 KearneyClinical predictors in the use.pdf%5Cnpapers2://publication/uuid/2103A28A-6180-4735-969C-5A9F8519ED63.
41. Boyle P, Gould AL, Roehrborn CG. Prostate volume predicts outcome of treatment of benign prostatic hyperplasia with finasteride: meta-analysis of randomized clinical trials. *Urology* 1996;**48**(3):398–405. [Internet]. Available from: http://www.ncbi.nlm.nih.gov/pubmed/8804493.
42. Kaplan SA, Lee JY, Meehan AG, Kusek JW. Long-term treatment with finasteride improves clinical progression of benign prostatic hyperplasia in men with an enlarged versus a smaller prostate: data from the MTOPS trial. *J Urol* 2011;**185**(4):1369–73.
43. Roehrborn CG, Lukkarinen O, Mark S, Siami P, Ramsdell J, Zinner N. Long-term sustained improvement in symptoms of benign prostatic hyperplasia with the dual 5alpha-reductase inhibitor dutasteride: results of 4-year studies. *BJU Int* 2005;**96**(4):572–7.
44. Gittelman M, Ramsdell J, Young J, McNicholas T. Dutasteride improves objective and subjective disease measures in men with benign prostatic hyperplasia and modest or severe prostate enlargement. *J Urol* 2006;**176**(3):1045–50.
45. Nickel JC, Gilling P, Tammela TL, Morrill B, Wilson TH, Rittmaster RS. Comparison of dutasteride and finasteride for treating benign prostatic hyperplasia: the Enlarged Prostate International Comparator Study (EPICS). *BJU Int* 2011:388–94.
46. McConnell JD, Bruskewitz R, Walsh P, Andriole G, Lieber M, Holtgrewe HL, et al. The effect of finasteride on the risk of acute urinary retention and the need for surgical treatment among men with benign prostatic hyperplasia. *N Engl J Med* 1998;**338**(9):557–63. [Internet]. Available from: http://eutils.ncbi.nlm.nih.gov/entrez/eutils/elink.fcgi?dbfrom=pubmed&id=9475762&retmode=ref&cmd=prlinks%5Cnpapers2://publication/doi/10.1056/NEJM199802263380901%5Cnhttp://www.nejm.org/doi/abs/10.1056/NEJM199802263380901.
47. Gravas S, Bach T, Bachmann A, Drake M, Gacci M, Gratzke C, et al. Guidelines on the management of non-neurogenic male lower urinary tract symptoms (LUTS), incl. benign prostatic obstruction (BPO). *Eur Assoc Guidel* 2015:1–70. [Internet]. Available from: https://uroweb.org/wp-content/uploads/EAU-Guidelines-Non-Neurogenic-Male-LUTS-Guidelines-2015-v2.pdf.
48. Mcvary KT, Roehrborn CG, Avins AL, Barry MJ, Bruskewitz RC, Donnell RF, et al. Update on AUA guideline on the management of benign prostatic hyperplasia. *J Urol* 2011;**185**:1793–803.
49. McConnell JD, Roehrborn CG, Bautista OM, Andriole GL, Dixon CM, Kusek JW, et al. The long-term effect of doxazosin, finasteride, and combination therapy on the clinical progression of benign prostatic hyperplasia. *N Engl J Med* 2003;**349**(25):2387–98. [Internet]. Available from: http://www.nejm.org/doi/abs/10.1056/NEJMoa030656.
50. Roehrborn CG, Barkin J, Tubaro A, Emberton M, Wilson TH, Brotherton BJ, et al. Influence of baseline variables on changes in International Prostate Symptom Score after combined therapy with dutasteride plus tamsulosin or either monotherapy in patients with benign prostatic hyperplasia and lower urinary tract symptoms: 4-year results. *BJU Int* 2014;**113**(4):623–35.
51. Yeo JK, Choi H, Bae JH, Kim JH, Yang SO, Oh CY, et al. Korean clinical practice guideline for benign prostatic hyperplasia. 2016;30–44.
52. Giuliano F, Ückert S, Maggi M, Birder L, Kissel J, Viktrup L. The mechanism of action of phosphodiesterase type 5 inhibitors in the treatment of lower urinary tract symptoms related to benign prostatic hyperplasia. *Eur Urol* 2013:506–16.

53. Rees J, Bultitude M, Challacombe B. The management of lower urinary tract symptoms in men. *BMJ* 2014;**348**:g3861. [June 24, 2015, Internet]. Available from: http://www.ncbi.nlm.nih.gov/pubmed/24963071%5Cnhttp://www.bmj.com/cgi/doi/10.1136/bmj.g3861.
54. Sairam K, Kulinskaya E, McNicholas TA, Boustead GB, Hanbury DC. Sildenafil influences lower urinary tract symptoms. *BJU Int* 2002;**90**(9):836–9.
55. Mulhall JP, Guhring P, Parker M, Hopps C. Assessment of the impact of sildenafil citrate on lower urinary tract symptoms in men with erectile dysfunction. *J Sex Med* 2006;**3**(4):662–7. [Internet]. Available from: http://www.ncbi.nlm.nih.gov/pubmed/16839322.
56. McVary KT, Roehrborn CG, Kaminetsky JC, Auerbach SM, Wachs B, Young JM, et al. Tadalafil relieves lower urinary tract symptoms secondary to benign prostatic hyperplasia. *J Urol* 2007;**177**(4):1401–7. [Internet]. Available from: http://www.ncbi.nlm.nih.gov/pubmed/17382741%5Cnhttp://ac.els-cdn.com/S0022534706030710/1-s2.0-S0022534706030710-main.pdf?_tid=0ef2a854-4235-11e3-9b73-00000aacb35d&acdnat=1383228320_542a9479767e726d361734c31ebbb905.
57. Porst H, McVary KT, Montorsi F, Sutherland P, Elion-Mboussa A, Wolka AM, et al. Effects of once-daily tadalafil on erectile function in men with erectile dysfunction and signs and symptoms of benign prostatic hyperplasia. *Eur Urol* 2009;**56**(4):727–36.
58. Stief CG, Porst H, Neuser D, Beneke M, Ulbrich E. Benign prostatic enlargement a randomised, placebo-controlled study to assess the efficacy of twice-daily vardenafil in the treatment of lower urinary tract symptoms secondary to benign prostatic hyperplasia. *Eur Urol* 2008;**53**:1236–44.
59. Roehrborn CG, McVary KT, Elion-Mboussa A, Viktrup L. Tadalafil administered once daily for lower urinary tract symptoms secondary to benign prostatic hyperplasia: a dose finding study. *J Urol* 2008;**180**(4):1228–34.
60. Oelke M, Giuliano F, Mirone V, Xu L, Cox D, Viktrup L. Monotherapy with tadalafil or tamsulosin similarly improved lower urinary tract symptoms suggestive of benign prostatic hyperplasia in an international, randomised, parallel, placebo-controlled clinical trial. *Eur Urol* 2012;**61**(5):917–25.
61. Glina S, Roehrborn CG, Esen A, Plekhanov A, Sorsaburu S, Henneges C, et al. Sexual function in men with lower urinary tract symptoms and prostatic enlargement secondary to benign prostatic hyperplasia: results of a 6-month, randomized, double-blind, placebo-controlled study of tadalafil coadministered with finasteride. *J Sex Med* 2015;**12**(1):129–38. [Internet]. Available from: http://www.sciencedirect.com/science/article/pii/S1743609515309073.
62. Casabé A, Roehrborn CG, Da Pozzo LF, Zepeda S, Henderson RJ, Sorsaburu S, et al. Efficacy and safety of the coadministration of tadalafil once daily with finasteride for 6 months in men with lower urinary tract symptoms and prostatic enlargement secondary to benign prostatic hyperplasia. *J Urol* 2014;**191**(3):727–33. [Internet]. Available from: http://www.sciencedirect.com/science/article/pii/S0022534713055584.
63. Füllhase C, Chapple C, Cornu J-N, De Nunzio C, Gratzke C, Kaplan SA, et al. Platinum priority – collaborative review – benign prostatic hyperplasia systematic review of combination drug therapy for non-neurogenic male lower urinary tract symptoms. *Eur Urol* 2013;**64**(2):228–43. [Internet]. Available from: https://doi.org/10.1016/j.eururo.2013.01.018.
64. Mansfield KJ, Liu L, Mitchelson FJ, Moore KH, Millard RJ, Burcher E. Muscarinic receptor subtypes in human bladder detrusor and mucosa, studied by radioligand binding and quantitative competitive RT-PCR: changes in ageing. *Br J Pharmacol* 2005;**144**(8):1089–99. [Internet]. Available from: http://doi.wiley.com/10.1038/sj.bjp.0706147.
65. Mukerji G, Yiangou Y, Grogono J, Underwood J, Agarwal SK, Khullar V, et al. Localization of M2 and M3 muscarinic receptors in human bladder disorders and their clinical correlations. *J Urol* 2006;**176**(1):367–73. [Internet]. Available from: http://www.ncbi.nlm.nih.gov/pubmed/16753445.

66. Lee JY, Kim HW, Lee SJ, Koh JS, Suh HJ, Chancellor MB. Comparison of doxazosin with or without tolterodine in men with symptomatic bladder outlet obstruction and an overactive bladder. *BJU Int* 2004;**94**(6):817–20.
67. Liao CH, Kuo YC, Kuo HC. Predictors of successful first-line antimuscarinic monotherapy in men with enlarged prostate and predominant storage symptoms. *Urology* 2013;**81**(5):1030–3.
68. Kaplan SA, Roehrborn CG, Rovner ES, Carlsson M, Bavendam T, Guan Z. Tolterodine and tamsulosin for treatment of men with lower urinary tract symptoms and overactive bladder: a randomized controlled trial. *J Am Med Assoc* 2006;**296**(19):2319–28. [Internet]. Available from: http://www.ncbi.nlm.nih.gov/pubmed/17105794.

Eastern Herbal Medicine

4

Dominique Thomas, Bilal Chughtai
Weill Cornell Medicine, New York, NY, United States

Introduction

For more than 2000 years, the use of Chinese herbal medicine has been practiced as a means to heal the body from symptoms and disease. During this time period, Chang Chung-Ching created a comprehensive guide known as the *Shang Han Lun* or "Treatise on Febrile Diseases," which is considered the first text detailing the use of Chinese medicine. Kanpo, also known as the "the way of Han," is composed of different ancient formularies, which were originally from Japan were introduced to Chinese culture during the Sui and Tang Dynasties. The total number of herbs currently used ranges from 5 to 8. Similarly, another text known as *Chin Kuei Yao Lueh* or "Prescriptions from the Golden Chamber" is comprised of ancient formularies, which is currently prescribed similarly to how it was in ancient times. Historically, the specific dosing for the different herbal formulas used in Kanpo is based on clinical efficacy.

Chinese Herbal Theory

The underlying philosophy of Chinese medical theory is known as the ***Yin–yang*** theory. Similar to the *Shang Han Lun*, it was developed over 5000 years ago. Yin is known to be the dark, passive, and deep aspects of medicine, while Yang takes on the active, strong, and light characteristics.[1] According to the Chinese medical theory, all conditions and diseases are the composition of imbalances of both yin and yang. When this imbalance occurs, if the person is exhibiting symptoms of too much yin, the patient would be given yang tonics and vice versa. Furthermore, disease in the yin–yang theory can be classified as being related to a fever or chill, interior or surface, and strong or weak.

During the 18th century, the **Qi, Blood, and Water** theory was developed in Japan. A person's Qi or Ki ("Chee") is known as the life force that runs throughout the body, which has also being described as the body's energy or vitality. Qi is synonymous with yang. Bodily fluids and blood would be considered to be yin in Chinese medicine. When there is a balance between the Ki and these fluids, the body should not experience any illness. It is when there is imbalance between the three that illness arises.[1] These herbal formulas of Qi are usually treated depending on the theory mentioned above.

When there is a blockage of blood flow, this is known Oketsu, which is an important diagnostic tool used in the practice of Kanpo. To properly locate Oketsu in the

A Comprehensive Guide to the Prostate. https://doi.org/10.1016/B978-0-12-811464-3.00004-0

body, herbalists must palpate the abdomen around the different points of the navel. When performing this, if the patient experiences pain, this is where the Oketsu is suspected to be located. Oketsu is thought to be the cause chronic inflammation, which causes blockages to the blood flow.[2]

Additionally, the properties of the herbal formula is very important when using a specific formula Within these practices there are currently 10 different formulas currently utilized, which include contracting, drying, moistening properties, etc. Depending on the condition, a patient would be given the opposite combination to treat the condition. For example, if a person presents with a fever, they would be given a cooling combination. Lastly, the origin of herbs is very complex in nature for they can have many different properties.[1]

Diagnostic Procedures

As a diagnostic procedure, Chinese herbal medicine has been used by many healers over thousands of years before the adoption of more modern day procedures. The most important method used in the diagnosis of patients has been the herbalist's sense. This means relying on what they see, smell along with feeling, and asking questions. In the modern day era, this is known as *Four Examinations*.

Four Examinations

1. **Visual:** This is the most important aspect of diagnosing the patient. Herbalists should be keen on observing how the patient presents and what the different features such as eyes and nose look like.
2. **Listening:** This is the second aspect of the examination. This includes using sound and what the patient is saying to make a proper diagnosis. It is important to pay particular attention to how the patient describes their condition along with how the patient sounds.
3. **Smell:** This is the third tool used during the examination. Herbalists should observe how the patient smells and if there are any odors.
4. **Palpation:** The last of the examination includes touch. The main aspect of this includes diagnosing the patient through palpating the abdomen. In total there are 14 different Kanpo abdominal areas of focus. Some are known for their simplicity such as emptiness below the navel, while others are much more complicated. These confirmations act as a guide for the herbalist to create a specific treatment regimen of herbal formulas.

Modern Research

Over the last few decades, there have not been many clinical research studies exploring the safety and efficacy of Chinese herbal medicines. However, physicians located in parts of Asia have been motivated to prove these formularies are sound methods of treating conditions. A main issue is that research studies conducted in Japan and China are oftentimes not translated into English, leaving them inaccessible to many parts of the Western world. Although there is a paucity of raw data detailing these properties in English, many herbalists in Western practice are confident in the use of these herbs given their survival over thousands of years. Furthermore, those practicing Kanpo are keen on its ability to treat the patient without causing too many side effects.[3] One of the many paradigms of Kanpo, which overshadows the use of pharmaceutical drugs, is this treatment is tailored to patients. Each patient is treated as an individual according to their symptoms and diagnosis. The sentiment is Kanpo will last the test of time despite new strains of bacteria and new diseases.

In a review by Ma et al., researchers evaluated the safety and efficacy of Chinese herbal medicine for BPH in comparison to Western medicines.[4] A total of 31 studies were included in their analysis. Chinese herbal medicine when prescribed alone or in tandem with Western medicine was shown to have a similar effect to placebo or when treating BPH patients with Western medicines. However, this study further demonstrated that many of the studies were poorly designed with very small patient populations. Thus meaning that the present studies do not completely support the use of herbal medicine when treating these patients.

Benign Prostatic Hyperplasia

Men with complaints of BPH have been known to be dealing with Oketsu. In Chinese medicine, this is characterized by constriction of blood flow to the prostate. Patients experiencing BPH have a multitude of symptoms such as pain, dysfunction of urination, and a decrease in quality of life. Combinations, including cinnamon and hoelen and persica and rhubarb, have shown some clinical efficacy across eastern medicine practices. Other symptoms include erectile dysfunction, urinary retention, and pelvic pain. A combination of rehmannia or polyporus has shown improvement in these symptoms. These herbs are responsible for targeting the tissues involved in inflammation. Less commonly, patients with dark urine or itching in this region usually improve with a combination of Gentiana.

Lower Urinary Tracts Symptoms

When it comes to lower urinary tract symptoms (LUTS), Chinese herbal medicine is effective in treating men experiencing stress or urge incontinence. Men with nocturia or incontinence during sleeping hours a combination of cinnamon and peony, Rehmannia 8, lotus seed, or Pueraria may be used. These herbs are equipped to address the abdominal muscles and pelvic floor muscles to help with atrophy or tightness. If the patient has no previous information specifying sensitivity to urge incontinence, the best herb to utilize is rehmannia 8 combination. When the patient presents with stress incontinence, the best combinations are Minor Bupleurum, Minor Cinnamon, and Peony Combination.

Conclusion

Chinese medicine and *Kanpo* in particular have much to offer in the area of urological conditions such as BPH and LUTS. Chinese herbs can help many patients to improve their conditions. While more research needs to be done, Chinese herbs have thousands of years of empirical efficacy and offer the patient another potential path toward a healthier state of being.

References

1. Kuwaki T. *Chinese herbal therapy: a guide to its principles & practice*. Long Beach (CA): Oriental Healing Arts Institute; 1990.
2. Dawes N. *Class notes*. Soho Health Arts; 2003.
3. Rister R. *Japanese herbal medicine: the healing art of Kampo*. Avery; 1999.
4. Ma CH, Lin LSL, et al. Efficacy and safety of Chinese herbal medicine for benign prostatic hyperplasia: systematic review of randomized controlled trials. *Asian J Androl* 2013;**15**:471–82.

Acupuncture

5

Dominique Thomas, Bilal Chughtai
Weill Cornell Medicine, New York, NY, United States

Benign Prostatic Hyperplasia

Benign prostatic hyperplasia (BPH) is a chronic condition affecting anywhere from 50% to 80% of aging men.[1] Because of the proliferation of prostate growth, there are a host of sequelae such as lower urinary tract symptoms (LUTS), including urinary incontinence.[2] There is currently an armamentarium of treatments available for men with these conditions ranging from medical therapy to surgery. Some patients may opt for more alternative methods in treating their symptoms. Acupuncture is type of medicine that utilizes thin needles, which are inserted into different sections of the body stimulating the nerves. Similar to Eastern medicine, the practice of acupuncture believes that the qi "the life force" of the body is circulating about parts of the body known as meridians. The insertion of needles about these channels is thought to restore any imbalance experienced by the patient.[3] The main purpose of this practice is to relieve discomfort.

A study by Wang et al. evaluated the use of electroacupuncture (EA) for the treatment of moderate and severe BPH at established acupoints and nonacupoints. A total of 100 patients between the ages of 50–70 with an International Prostate Symptom Score (IPSS) ≥8 were included in this study. Over a 4-week period, there were a total of 16 treatments given to each patient. The total duration of the study was 6 weeks with follow-up up to 18 weeks. The primary effectiveness outcome was a difference in IPSS at the first follow-up time period. Those who were treated at EA acupoints showed a "4.51 ($p<0.001$) and 4.12 ($p<0.001$) points greater decline in IPSS than the control."[4] By the last follow-up point at 18 weeks there was a statistically significant decrease in IPSS by 3.2 points ($P=.001$). In comparison to controls, there was not a significant reduction in Qmax ($P=.819$) or PVR ($P=.35$) at the 6 weeks.[4] By the end of the 18 weeks, those in the treatment arm had improvements that remained durable.

Another study by Yu et al. looked at the effects of EA in patients with LUTS secondary to BPH. A total of 42 patients were randomized into the trial receiving either "2 Hz EA for 20 min twice/week for a total of twelve treatments, or a sham EA group (CG), received sham EA."[5] Baseline demographics showed these patients were over the age of 40, had an IPSS >8 and a Qmax <15 mL/s. The primary outcome was improvements in Qmax, average flow rate, voided volume, and total flow time after treatment. Those in the EA group showed "voiding volume, average flow rate, and maximal flow rate in the EG were 32.2 ± 104.4 mL, 1.2 ± 1.6mL/sec, and 2.3 ± 3.7 mL/sec, respectively."[5] In the sham group, there were increases in their measures by "−37.9 ± 120.4, −0.22 ± 2.7, and −0.3 ± 4.3, respectively ($P=.038$, .026, and .030)."[5] Lastly there were no significant changes in prostate-specific antigen or IPSS in comparison to the two groups

A Comprehensive Guide to the Prostate. https://doi.org/10.1016/B978-0-12-811464-3.00005-2

($P=.573$, .175). These data demonstrate the clinical effectiveness of the 2 Hz EA was limited in its ability to improve symptoms related to LUTS secondary to BPH.

To date, there have not been many high-quality randomized controlled trials (RTCs) evaluated the effectiveness of acupuncture for BPH. A recently published systematic review and metaanalysis assessed the therapeutic effects of this treatment in RCTs. A total of eight RTCs were included in this analysis with a total number of 661 men with BPH.[6] The follow-up in this study ranged from 4 weeks to 18 months. Three trials compared the results of acupuncture to a control group and had relatively short follow-up (4–6 weeks). These trials demonstrated during this follow-up period, there was a significant improvement in IPSS (MD −1.90, 95% CI −3.58 to −0.21).[6] Clinically, there was a minimal clinical important difference. Those trials with moderate follow-up (n = 12, 12–18 weeks) showed no statistically significant effect on IPSS (MD -2.04, 95% CI −4.19, 0.10).[6]

The effects of acupuncture need to be better elucidated in the literature through the use of longer-term trials.

Conclusion

BPH is a chronic urological disorder effecting the aging male population. Preliminary results have shown acupuncture maybe an effective method in treating those who are symptomatic. Furthermore, this technique can be used alone or in combination with a multidisciplinary treatment strategy. Although there is some research addressing its effectiveness, the current literature is lacking with long-term effects of it durability in this cohort. Well-designed RTCs evaluating its effectiveness to other treatments are needed along with an understanding of the mechanism of action which acupuncture alleviates symptoms. Of the studies that are available, acupuncture has a relatively low adverse effect profile and can be tried of those who do not have an absolute need of a surgical intervention.

References

1. Chughtai B, Forde JC, Thomas DD, et al. Benign prostatic hyperplasia. *Nat Rev Dis Primers* 2016;2:16031.
2. McVary KT, Roehrborn CG, Avins AL, et al. Update on AUA guideline on the management of benign prostatic hyperplasia. *J Urol* May 2011;**185**(5):1793–803.
3. *Acupuncture – mayo clinic*. 2017.
4. Wang Y, Liu B, Yu J, Wu J, Wang J, Liu Z. Electroacupuncture for moderate and severe benign prostatic hyperplasia: a randomized controlled trial. *PLoS One* 2013;**8**(4):e59449.
5. Yu JS, Shen KH, Chen WC, Her JS, Hsieh CL. Effects of electroacupuncture on benign prostate hyperplasia patients with lower urinary tract symptoms: a single-blinded, randomized controlled trial. *Evid Based Complement Altern Med* 2011;**2011**.
6. Zhang W, Ma L, Bauer BA, Liu Z, Lu Y. Acupuncture for benign prostatic hyperplasia: a systematic review and meta-analysis. *PLoS One* 2017;**12**(4):e0174586.

Naturopathic and Western Medicine

6

Dominique Thomas, Zobia Shah, Bilal Chughtai
Weill Cornell Medicine, New York, NY, United States

Naturopathic Herbal Medicine for Benign Prostatic Hyperplasia

There are a myriad of ways to treat benign prostatic hyperplasia (BPH) ranging from watchful waiting to surgical intervention. Another method known as Naturopathic and Western medicine combines the use of herbal treatments and the comprehension of entire body to treat the root cause of this disease. When relating to BPH, utilizing Western herbs can be distributed into two distinct groups. These include those methods that are used to treat the symptoms experienced by the patient, which is distinct from those specifically targeting the cause of the disease.

Although we have these categories some herbal treatments function in tandem to treat the patient holistically. Serenoa repens fruit also known as saw palmetto is currently one of the most researched western herbs. This herb is currently harvested for use in Florida and is native to the southeastern parts of the United States and the Caribbean. Beginning in the 1950s, saw palmetto has been the subject of many clinical trials. Researchers have studied the mechanism of action of saw palmetto showing different components of the herb are responsible for acting on pathways within the prostate. The synergistic effects of these constituents lead to an overall clinical effect. Despite this fact, this does not have the same effects in all patients.[1] Saw palmetto is composed of an array of phytosterols and fatty acids, which are responsible for nonspecific 5α-reductase inhibition and calcium channel blockers.[1] When compared to other medical therapies such as finasteride, saw palmetto is effective in inhibiting type 1 and 2 5α-reductase by 32%–50%, while finasteride has an inhibition rate of 80% for type 2.[1] Because of this patients taking saw palmetto experience less adverse effects while many patients taking finasteride may experience low libido and sexual dysfunction. A caveat to this information is saw palmetto is not effective in decreasing the size of the prostate.

In 2012, MacDonlad et al. conducted a metaanalysis of saw palmetto extracts in the treatment of BPH.[2] The main purpose of this study was to evaluate the efficacy of patients taking saw palmetto alone in comparison to a placebo.[2] At a total of 17 (N=2008) randomized controlled trials were included in this study. Demographic characteristics showed the average age of patients was 64.3 years with the majority of those being Caucasian. The length of follow-up ranged from 4 to 72 weeks. The majority of patients had moderately severe symptoms as indicated with a baseline score of 14 on the American Urological Association Symptom Index (AUASI). Of the studies reported

A Comprehensive Guide to the Prostate. https://doi.org/10.1016/B978-0-12-811464-3.00006-4

a total of three trials had long-to-moderate results. Based on these trials (n=661) it was found saw palmetto was no different when compared to placebo when assessing symptoms associated with LUTS through the use of the AUASI and maximum flow rate (Qmax). Furthermore, there was no difference between treatment groups when comparing Qmax in those in short-term studies. In sum, despite assumptions saw palmetto is effective in treating symptoms of BPH, this review demonstrated that it was not effective when compared to placebo in improving AUASI, Qmax, or nocturia ($P=.19$).[2]

Another study by Carraro et al. randomized a total of 1098 men with mild-to-moderate symptoms of BPH to receive either saw palmetto or finasteride with prostates under 50cc.[3] Subjects taking saw palmetto had a lower adverse effect profile when compared to those on finasteride. Furthermore, saw palmetto "and finasteride decreased the IPSS (−37% and −39%, respectively), improved quality of life (by 38% and 41%), and increased peak urinary flow rate (+25% and +30%, $P=.035$)."[3] Saw palmetto was not effective in changing the size of the prostate or effecting PSA levels.

Given the results of the two studies mentioned above, interesting saw palmetto was as effective as finasteride, a drug used to treat symptoms of BPH but in contrast was less effective as placebo. This may be due to bias in the design of the trial. Saw palmetto can be used as an effective measure of reducing symptoms of BPH. It is inexpensive and is relatively safe for those taking the extract. Furthermore, research studies have shown some clinical efficacy for those who take the recommended dose of 320mg/day containing roughly 70%–90% fatty acids.[4]

Another herbal treatment used for BPH symptoms, which is a similar profile to saw palmetto, is *Prunus africanum* (pygeum). Despite this it is not approved for use in North American countries because removing the bark from these trees leads to tree death. Furthermore, the vast population of these trees tends to grow in areas of the African continent.[5] To date there have been a paucity of research evaluating the clinical effects of this treatment option. A Cochrane review found a total of 18 randomized trials with 1562 who were treated with pygeum. However, there were no trials assessing it in comparison to other pharmacologic agents.[6] The mean length of follow-up was 64days. Pygeum was effective in largely improving urinary symptoms when compared to placebo. Nocturia in this group was reduced by 19%, Qmax was increased by 23%, and residual urine volume by 24%.[6]

Urtica dioica (nettle) root is also known to be quite effective in treating men with mild-to-moderate symptoms of BPH. This root binds to sex hormones inhibiting synthesis. Like other nutraceuticals, there have not been very many large trials showing its effectiveness. In a study by Safarinejad, evaluated the use of nettle root during a 6-month randomized placebo-controlled trial. A total of 558 patients completed the study.[7] Those in the treatment arm reported an 81% improvement in LUTS symptoms in comparison to 16% in the placebo arm ($P<.001$). Furthermore, those taking nettle root showed marked improvement in IPSS ($P=.002$), peak flow rates ($P<.05$), and postvoid residual volume ($P<.05$).[7] Those taking placebo did not show any significant changes. Nettle root was beneficial for men with symptomatic BPH.[7] Interestingly, when combining the efficacy of saw palmetto and nettle root men with symptomatic BPH had superior effects to those taking finasteride alone when dosing at 500–1000mg.[8]

Conclusion

Naturopathic western medicine has shown some promise in men with symptomatic BPH. These herbal treatments similar Eastern medicine have been around for thousands of years. These treatments are relatively safe with a low adverse effect profile. Long-term clinical trials are needed to assess the role of these treatments in comparison to other pharmaceutical interventions.

References

1. Koch E. Extracts from fruits of saw palmetto (*Sabal serrulata*) and roots of stinging nettle (*Urtica dioica*): viable alternatives in the medical treatment of benign prostatic hyperplasia and associated lower urinary tracts symptoms. *Planta Med* August 2001;**67**(6):489–500.
2. MacDonald R, Tacklind JW, Rutks I, Wilt TJ. *Serenoa repens* monotherapy for benign prostatic hyperplasia (BPH): an updated Cochrane systematic review. *BJU Int* June 2012;**109**(12):1756–61.
3. Carraro JC, Raynaud JP, Koch G, et al. Comparison of phytotherapy (Permixon) with finasteride in the treatment of benign prostate hyperplasia: a randomized international study of 1098 patients. *Prostate* October 1996;**29**(4):231–40. discussion 241–232.
4. *Saw palmetto for enlarged prostate*. 2017.
5. Kate K, Laird SA. *The commercial use of biodiversity*. 1999.
6. Wilt T, Ishani A, Mac Donald R, Rutks I, Stark G. *Pygeum africanum* for benign prostatic hyperplasia. *Cochrane Database Syst Rev* 2002;(1):Cd001044.
7. Safarinejad MR. *Urtica dioica* for treatment of benign prostatic hyperplasia: a prospective, randomized, double-blind, placebo-controlled, crossover study. *J Herb Pharmacother* 2005;**5**(4):1–11.
8. Sokeland J, Albrecht J. [Combination of *Sabal* and *Urtica* extract vs. finasteride in benign prostatic hyperplasia (Aiken stages I to II). Comparison of therapeutic effectiveness in a one year double-blind study]. *Urol A* July 1997;**36**(4):327–33.

Monopolar Transurethral Resection of Prostate

7

Niall F. Davis, James C. Forde
Beaumont Hospital, Co Dublin, Ireland

Introduction

Monopolar transurethral resection of prostate (TURP) has been the primary surgical treatment option for lower urinary tract symptoms (LUTS) due to benign prostatic hyperplasia (BPH) for >80 years.[1] The prevalence of benign prostatic enlargement and associated LUTS increases with age and approximately 25% of men have received treatment for LUTS by the age of 80 years.[2,3] This increasing prevalence of LUTS in an aging global population has prompted clinicians to pursue alternative urological surgical modalities to replace monopolar TURP. However, throughout the current technologically innovative urological era monopolar TURP has remained the standard reference point for surgical treatment of BPH due to the availability of robust long-term functional data that describe efficacy and morbidity.[4] No other treatment has surpassed monopolar TURP in terms of symptomatic improvement and durability.[5] Furthermore, recent developments in improved anesthesia and advances in surgical technique and instrumentation over the last 30 years have ensured that monopolar TURP remains the reference standard urological procedure for assessing surgical outcomes after treatment of LUTS due to BPH.[6]

Monopolar TURP involves monopolar diathermy, which entails an electrical current flowing from an active electrode through adenomatous prostatic tissue and the body before exiting via a return skin electrode.[7] Hypoosmolar irrigation fluids such as glycine are required intraoperatively as they are molecularly inert, optically clear and nonconductive. Water is discouraged for irrigation purposes due to the risk of hemolysis.[8] Irrigating solutions predispose the patient to dilutional hyponatremia and transurethral resection (TUR) syndrome, which is characterized by metabolic abnormalities, cardiovascular compromise, and neurological dysfunction. TUR syndrome has a mortality rate of up to 40%.[9] Importantly, better patient selection, technological advances, and increased awareness of potential complications have resulted in a considerable reduction in morbidity and mortality from monopolar TURP.[1]

Historical Perspective

The concept of TURP was pioneered in the United States in the 1920s and many important historical factors at the time were significant for prompting its initial development into the field of urology[10,11]:

1. Invention of the lamp by Edison in 1879
2. Development of the cystoscope in 1887

A Comprehensive Guide to the Prostate. https://doi.org/10.1016/B978-0-12-811464-3.00007-6

3. Development of the fenestrated tube that allowed the obstructing tissue to be blindly excised
4. Invention of the vacuum tube in 1908 that allowed a continuous production of high-frequency electrical current for resecting tissue.

In 1926, the cystoscope was combined with the tubular punch and a tungsten loop was developed to enable tissue resection. McCarthy assembled this group of instruments together in 1932 with an oblique lens so that resection under direct vision was feasible instead of the blind procedures that were performed up to this point.[12,13] Fiberoptic lighting systems were developed in the 1970s, which significantly improved endoscopic visualization. The rod–lens system, developed by Hopkins in 1976, further improved visualization as more light was transmitted through the lens system.[14,15]

Investigating Lower Urinary Tract Symptoms

All clinicians should take a thorough patient history and perform a clinical examination during the initial consultation in a patient with bothersome LUTS. An assessment of symptom severity using the International Prostate Symptom Score should also be performed.[16] There are considerable differences among guidelines about diagnostic investigations for investigating LUTS. High-quality European Association of Urology (EAU) and American Urological Association (AUA) guidelines that are developed based on randomized controlled trials only recommend a few tests such as urine analysis and a fully completed voiding diary and measurement of serum creatinine if renal impairment is suspected (Table 7.1).[17–19] Digital rectal examination (DRE) is always performed to investigate the presence of prostate cancer and to estimate prostate size.

Uroflowmetry and measurement of postvoid residual (PVR) volumes are considered optional during the initial assessment of LUTS but are a prerequisite prior to TURP as well as serum creatinine measurement (Table 7.2).[17,18,20] Prostate specific

Table 7.1 Recommended diagnostic investigations for lower urinary tract symptoms (LUTS)[17–19]

Investigations for LUTS	
Always	**Optional**
Medial history	PSA
Clinical examination with DRE	Uroflowmetry
Validated symptom score questionnaire	Postvoid residual
Frequency/volume chart	Ultrasonography
Urinalysis	Cystourethroscopy
	Pressure–flow studies
	Serum creatinine

DRE, digital rectal examination; *PSA*, prostate specific antigen.

Table 7.2 **Recommended diagnostic investigations prior to monopolar transurethral resection of the prostate (TURP)[17–19]**

Investigations prior to performing monopolar TURP	
Always	**Optional**
Medial history Clinical examination with DRE Validated symptom score questionnaire Frequency/volume chart Urinalysis Serum creatinine Uroflowmetry Postvoid residual Ultrasonography	PSA Cystourethroscopy Pressure-flow studies

PSA, prostate specific antigen.

antigen (PSA) measurement can be taken after appropriate counseling as a PSA level >1.4 ng/mL is an independent predictor of symptom progression. Transrectal ultrasonography may be performed if there is clinical suspicion of a prostate gland >100 mL as higher volumes increase the likelihood of open prostatectomy. Pressure flow studies are useful for predicting symptomatic outcomes after TURP; however, they are invasive to the patient and many clinicians believe their invasiveness outweighs their clinical benefit. Renal ultrasonography should be performed if serum creatinine levels are elevated to detect hydronephrosis associated with high-pressure chronic urinary retention. The percentage of patients with hydronephrosis on ultrasonography is directly proportional to their serum creatinine level: creatinine <115 mmol/L: 0.8%, creatinine 115–130 mmol/L: 9%, and creatinine >130 mmol/L: 33%.[21]

Indications for Monopolar Transurethral Resection of Prostate

The indications for monopolar TURP have changed during the last 20–30 years due to the increasing use of medical therapy for symptomatic BPH.[22] Alpha-adrenergic antagonists and 5-alpha reductase inhibitors are now frequently used as first-line management in isolation or in combination to treat LUTS secondary to BPH when conservative measures fail. Currently, the most common indication for monopolar TURP is moderate-to-severe LUTS secondary to BPH that is refractory to medical therapy (Table 7.3).[16] Other common indications include abnormal uroflowmetry parameters, recurrent acute urinary retention, increasing PVR volumes, bladder stones, hydronephrosis secondary to BPH, recurrent hematuria secondary to BPH and recurrent urinary tract infections (UTIs) secondary to BPH. One retrospective study reported that "failed medical therapy" as an indication for TURP increased from 36% in 1998 to 87% in 2008 (Table 7.4).[18]

Table 7.3 Indications for performing monopolar transurethral resection of prostate (TURP)

Indications for monopolar TURP
Recurrent or refractory urinary retention
Recurrent **U**TIs
Bladder **S**tones
Hematuria
Elevated creatinine
Symptom deterioration

Note the pneumonic "RUSHES" is highlighted in the bold font to highlight the indications for TURP.
UTIs, urinary tract infections.

Table 7.4 Symptoms associated with transurethral resection syndrome relative to serum sodium levels[35]

Sodium concentration (mmol/L)	Symptoms
130–135	Asymptomatic
120–130	Confusion and nausea
115–120	Seizures
<115	Coma

Perioperative Considerations

Discontinuing Antithrombotic Agents

The risk of peri- and postoperative bleeding in patients taking antithrombotic agents should be balanced against the risk of developing a thromboembolic event.[23] When aspirin is discontinued, most studies recommend stopping the agent 7–10 days preoperatively and recommencing the agent on the first postoperative day provided there are no hemostatic concerns.[24,25] Similarly, most clinicians discontinue clopidogrel 7–10 days preoperatively and recommence this anticoagulant on the first postoperative day if there are no bleeding concerns.[24,25] Warfarin is an anticoagulant that is usually discontinued 5 days preoperatively and restarted on the first postoperative day.[25]

Patients undergoing TURP are in a hypercoagulable state and have a 0.1%–0.2% risk of developing a pulmonary embolus.[26] Therefore, above knee thromboembolism stockings are recommended until the patient is discharged. Furthermore, compression boots are administered intraoperatively and patients are mobilized early postoperatively.

Anesthesia

Monopolar TURP is usually performed with the patient under general or spinal anesthesia. There appears to be no difference in perioperative blood loss, morbidity, or mortality when both anesthetic techniques are compared.[27] The UK National Prostatectomy Audit found that approximately 3% of patients undergoing monopolar TURP required >2 units of blood and therefore recommend a group and save serum sample in all patients undergoing TURP.[26]

Antibiotic Therapy

Preoperatively, culture proven UTIs are found in 8%–24% of patients undergoing TURP and this incidence increases in patients with indwelling urinary catheters.[28] In this clinical scenario, it is important that the infection is actively treated preoperatively. Previously, the necessity for administering perioperative antibiotic therapy was unclear; however, a number of reports from the 1980s and 1990s demonstrated that perioperative antibiotics reduce the incidence of postoperative urosepsis.[28,29] The AUA and EAU guidelines recommend that patients are administered intravenous antibiotic therapy at induction.[17–19] A first generation cephalosporin by itself or combined with the aminoglycoside gentamicin is typically recommended with some urologists continuing with oral antibiotic therapy until the patient's transurethral urinary catheter is removed during the postoperative period. Intravenous amoxicillin with intravenous gentamicin is recommended for patients with heart murmurs or prosthetic heart valves with an additional dose of oral amoxicillin 6 h later.

Overview of Technique

There is a multitude of cutting techniques for resecting an obstructing prostate gland during monopolar TURP.[7] All endoscopic techniques require good vision, clear identification of the surgical capsule, and anatomical landmarks.[11] In the event of hemorrhage, pulsatile arterial bleeding should be addressed first followed by venous hemorrhage.[30,31] Occasionally, arterial bleeding cannot be coagulated without resecting additional adenomatous tissue with the cutting resectoscope. An Ellick evacuation bulb is used to evacuate resected adenoma, and it is important to ensure that all resected prostatic chips have been removed from the bladder and urethra at the end of the procedure and that adequate hemostasis has been secured. A three-way Foley catheter is inserted into the bladder and gently irrigated to confirm unobstructed flow of urine and irrigation fluid. Gentle traction on the catheter can enhance hemostasis through a tamponade effect. Video cameras are now used routinely with the resectoscope as they distance the surgeon from blood products, enhance vision, and provide an excellent method for teaching techniques in endoscopic surgery.

Stepwise Technique

The patient is placed in the lithotomy position with the buttocks just off the end of the operating table and leg holders are used for support. The perineum is not shaved but scrubbed with a disinfectant and the patient is draped. A DRE is again performed to characterize the size and contour of the prostate gland. A rigid cystourethroscopy is also performed to identify anatomical landmarks, including the verumontanum the prostatic urethra, the bladder neck, and the ureteric orifices. The urethra is calibrated to 28 French (Fr) with urethral dilators to accommodate a 26 Fr or 28 Fr continuous flow resectoscope. If a prolonged resection is anticipated a suprapubic catheter can be inserted under direct vision to create a low-pressure continuous-flow circuit during the resection. After the initial cystourethroscopy and urethral calibration, the bladder is filled with 150 mL of nonhemolytic irrigation solution such as 1.5% glycine.[15]

Most operative techniques recommend resecting ventral adenomatous tissue initially so that dorsal adenoma can drop down facilitating stepwise resection under direct vision. Some clinicians recommend resecting the median lobe/floor of the prostate initially to create a channel or trough for improved flow of irrigation during the procedure. When the median lobe is prominent or intravesical we recommend resecting the median lobe at the beginning of the procedure and resecting the ventral tissue toward the end (Fig. 7.1).

1. The sheath of the resectoscope is usually controlled with the surgeon's nondominant hand at the level of the verumontanum. This fixed position prohibits the loop from resecting distal to the verumontanum. Resecting distal to the verumontanum can affect urinary continence postoperatively.
2. To resect the left lobe, develop the trough at 5 o'clock toward the 2 o'clock direction. Gradually deepen the resection down to the capsule of the prostate (Fig. 7.1B).

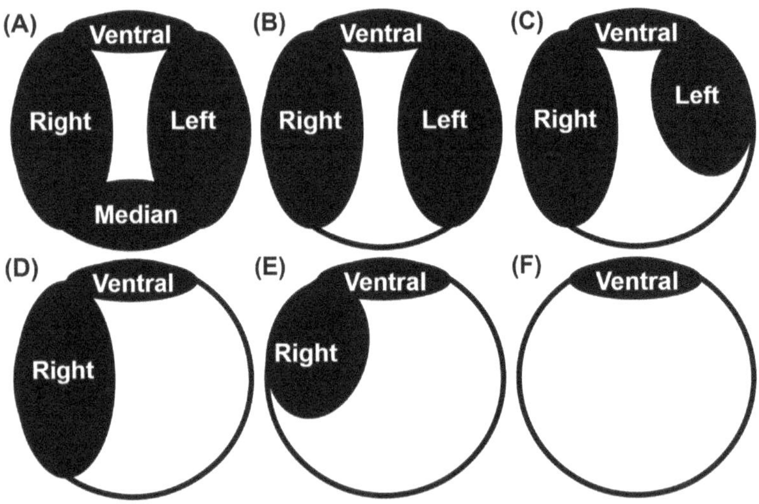

Figure 7.1 Stepwise resection during monopolar transurethral resection of the prostate. (A): The relevant anatomical landmarks and lobes of the prostate are identified before resection commences. (B): A prominent median lobe is resected at the beginning of the procedure. (C): To resect the left lobe, develop the trough at 5 o'clock towards the 2 o'clock direction. (D): Deepen the resection down to the capsule of the prostate and resect the remaining left lobe toward the 1 o'clock position. (E): The right lobe is resected by developing a trough from the 7 o'clock trough toward the 10 o'clock position. (F): The sheath is then rotated 180degrees to resect ventral adenomatous tissue from 10 o'clock to finish at 2 o'clock. Residual apical tissue can then be resected at the end of the procedure prior to securing hemostasis.

3. Develop the resected trench from 5 to 7 o'clock without undermining the bladder. Excessive resection or hemostasis with electrocautery can result in bladder neck stenosis due to ischemia with resultant fibrosis (Fig. 7.1D). Next, resect toward the 10 o'clock position.
4. To resect ventral adenomatous, tissue the sheath should be rotated 180 degrees at the level of the verumontanum and resect from 10 o'clock clockwise to finish at 2 o'clock (Fig. 7.1F).
5. Residual apical tissue can then be removed in a fully controlled manner by compact shorter resecting movements with the resectoscope loop. Any residual obstructing adenomatous tissue posterior to the bladder neck is resected after reidentifying both ureteric orifices.
6. Secure hemostasis with coagulating electrocautery and/or the roller ball. Discontinuing the irrigation temporarily is useful for identifying persistent venous bleeding and ooze.
7. Introduce a 22–24 French (Fr) three-way catheter and inflate the anchoring balloon with 30 mL of sterile water. Apply slight traction on the catheter to tamponade the prostatic cavity so that the balloon is anchored at the bladder outlet.
8. When the irrigation fluid becomes clear, catheter traction can be released after 4 h. Continue to irrigate the bladder overnight and the catheter can be removed 2 days postoperatively provided there are no early complications.

Commonly Encountered Intraoperative Issues

Several mechanical, patient or electronic intraoperative issues with equipment can occur during monopolar TURP. It is important to adopt a stepwise approach when analyzing and solving these problems. If the resectoscope loop does not cut the adenomatous prostatic tissue take the following investigatory approach[32]:

1. Ensure that the loop is correctly sitting in its holder by "clicking" into place.
2. Inspect the loop to ensure that it is not broken.
3. Examine the diathermy plate to check that it is securely attached to the thigh.
4. Check that the diathermy lead is securely attached to the diathermy machine.
5. Look at the irrigating fluid and confirm that it is glycine as opposed to saline.
6. If the loop still does not cut despite taking these measures, then change the diathermy machine.

Intraoperatively, if the clinician is unable to control bleeding vessels then the following stepwise measures should be taken[31]:

1. Compress the adenoma to one side of the artery so that it compresses the bleeding artery itself.
2. If the bleeding artery is straight at the lens then advance the resectoscope beyond the bleeder, angulate it to compress the artery and slowly withdraw the sheath until the open vessel is identified by a spurt of blood.
3. Look just inside the bladder neck as persistent bleeders are often found here.
4. For venous bleeding, pass a catheter into the bladder and instill 50 mL of water into the anchoring balloon to tamponade the bladder neck where most bleeding veins are located. Maintain traction for 10 min and do not remove any equipment in case the resectoscope needs to be reintroduced.

Complications

Over the past 70 years, there has been a significant reduction postoperative mortality associated with monopolar TURP. In the 1930s the mortality rate after 30 days was 5%.[11,21,26] Since the 1990s, the mortality rate after TURP has decreased to 0% after 30 days in many large retrospective studies.[33,34] Mebust et al. also reported on the incidence of significant complications after monopolar TURP and found that the most common early complications were failing to void (6.5%), bleeding requiring transfusion (3.9%), and clot retention (3.3%).[21]

Bleeding

The volume of intraoperative blood loss is dependent on several variables such as prostate size, duration of procedure, and level of surgical expertise.[31] Arterial bleeders are controlled by judicious and accurate use of electrocautery at each stage of the resection. If there is a persistent bright red color in the catheter bag despite irrigation at the end of the procedure the surgeon should have a high index of suspicion for arterial bleeding. Reinsert the resectoscope, washout any large clots and attempt to coagulate any arterial bleeding. Venous bleeding is characterized by a more delayed dark ooze from the catheter when it is irrigating and it is controlled by inflating 50 mL of fluid into the anchoring balloon and placing the catheter on traction for 10 min when the patient is still on the operating table. If conservative and endoscopic measures fail to control the bleeding, then the prostatic fossa should be packed. A lower midline or Pfannenstiel incision is performed and the bladder is opened. Packs are inserted around the urethral catheter for 48 h to tamponade any bleeding vessels. After 48 h, the patient should be returned to theater for removal of swabs and definitive closure of the anterior abdominal wall. Alternatively, uncontrollable bleeding can be managed with superselective internal iliac artery embolization if interventional radiology facilities are available.

Transurethral Resection Syndrome

The incidence of TUR syndrome is 2% in some reports and it typically presents with mental confusion/agitation, visual disturbances, nausea, vomiting, bradycardia, and hypertension.[35] Without treatment, circulatory disturbances, hyponatremic shock with kidney failure, brain, and lung edema can quickly develop. Patients typically become symptomatic when sodium levels drop below 125 mmol/L (Table 7.4).[35]

TUR syndrome is multifactorial and results from absorption of large volumes of glycine during monopolar TURP. The pathophysiology of TUR syndrome is due to dilutional hyponatremia, fluid overload, and glycine toxicity. Absorption of glycine occurs through the periprostatic venous plexus and through perivesical and retroperitoneal spaces. Ninety-percent of glycine is metabolized to ammonia, glycolic acid, and water by the liver.[35] The remaining 10% is metabolized by the kidney. Dilutional hyponatremia results in an osmotic shift of water from plasma across the blood–brain barrier and can lead to cerebral herniation and death is untreated. Glycine also induces an osmotic diuresis resulting in a further loss of sodium from the body. Finally, glycine itself is an inhibitory neurotransmitter in the retina and excessive concentrations lead to acute vision loss, bradycardia, and cardiotoxicity.[35]

Risk factors for developing TUR syndrome are a prostate gland is larger than 45 g and a resection time >90 min.[21] Initial management consists of administering intravenous frusemide (40 mg) to encourage diuresis and informing the HDU or ICU department. Cardiac arrhythmias should be treated according to cardiac guidelines. Patients with severe symptoms and signs of TUR syndrome should be managed in the intensive care unit, given 100% oxygen and sodium should be replaced when the serum sodium level drops below 120 mmol/L.[36] A correction of 1 mmol/L/h is recommended as rapid correction can result in central pontine myelinolysis.

The risk of TUR syndrome can be decreased by taking some important perioperative measures.[21] Firstly, the urologist should be aware of the patient's preoperative serum sodium level. Secondly, the hydrostatic pressure of the irrigating fluid should be limited to 50–60 cm water. Thirdly, the resection time should be limited to a maximum of 60 min with close patient monitoring and accurate documentation of the volume of irrigation fluid used throughout the procedure. If required, 2% volume ethanol can be added to the irrigating fluid and expiratory alcohol concentration levels can be measured with an alcometer.[21]

Perforation of the Capsule

Perforation of the prostatic capsule with extravasation occurs in approximately 2% of patients and is characterized intraoperatively by symptoms such as nausea, vomiting, and pain localized to the lower abdomen and back.[37] If perforation is suspected the operation should be discontinued emergently once hemostasis has been achieved. Over 90% of patients with capsular perforation are safely managed conservatively with an indwelling transurethral or suprapubic urinary catheter.[37]

Sepsis

The risk of urinary sepsis is 3% and the risk of systemic sepsis after monopolar TURP is approximately 1.5%.[37]

Late Complications

Reoperations or urethral dilatations due to urethral stricture disease or bladder neck stenosis occur in approximately 3%–5% of patients after monopolar TURP.[34] There is also a risk that the procedure itself will need to be repeated as the cumulative risk of repeat TURP is 2% per year.[34] The incidence of urinary incontinence after 12 months is <1%.[34] Erectile dysfunction deteriorates in approximately 17% of patients, improves in 28% and remains unchanged in 55%.[38] A decrease in international index of erectile function (IIEF) scores after monopolar TURP is usually transient with IIEF scores beginning to improve 3 months postoperatively. This improvement is maintained at 6 and 12 months postoperatively.[38]

Clinical Outcomes

The chance of improvement in a patient's preceding voiding LUTS after TURP ranges from 70% to 96% with a reduction in symptom score of 85%.[15,39] In the UK National Prostatectomy Audit on 5361 patients undergoing monopolar TURP, it was demonstrated that patients who were most bothered by their symptoms had the best response to surgery.[26] Elderly men and men of a higher social class were more likely to undergo monopolar TURP with fewer symptoms.[26] One study followed up a series of patients undergoing monopolar TURP for 3 years. The authors noted that 84% of patients reported an improvement in their symptoms at 12 months with 10% of patients remaining the same.[40] After 3 years, 75% maintained their improvement and 13% remained the same. Patients with a deterioration in their symptoms typically developed de novo urgency and urge incontinence.[40] Only one patient required a repeated monopolar TURP and 10% developed a bladder neck contracture. Another large retrospective study from the 1980s reported that approximately 90% of patients were satisfied with their clinical outcomes 5 years after monopolar TURP.[41] Objective flow rates improved from preoperative values.

Monopolar Versus Bipolar Transurethral Resection of Prostate

Several prospective randomized trial studies compared the safety and efficacy of monopolar TURP with bipolar TURP.[42] Although monopolar TURP is still regarded as the reference standard, it is currently challenged by a multitude of endoscopic transurethral techniques. Available evidence suggests that bipolar TURP is an attractive alternative to monopolar TURP as both urological techniques have comparable medium-term efficacy.[1,43] There are no significant differences between monopolar and bipolar in terms of clinical effectiveness, except for the results for Qmax where bipolar TURP is associated with a slightly higher Qmax.[44] Rieken et al. reported a significant and persistent improvement of voiding parameters and micturition symptoms that were comparable between both techniques after 5 years.[45] Monopolar TURP is associated with more significant morbidity compared to bipolar TURP with one systematic review demonstrating more adverse events with monopolar TURP Specifically, TUR syndrome occurred in 35/1375 (2.5%) patients undergoing monopolar TURP compared to no patients with bipolar TURP.[46]

Conclusions

Monopolar TURP was initially pioneered in urology as it was considered minimally invasive and less morbid compared to the traditional open prostatectomy. At present, monopolar TURP remains the reference endoscopic surgical treatment for male patients with LUTS due to BPH. The drive to create novel minimally invasive treatments arises from the desire to decrease perioperative, short-term, and long-term morbidities that are associated with monopolar TURP. Although recently developed technologies such as bipolar TURP and laser enucleation of the prostate demonstrate favorable perioperative outcomes with encouraging short- and medium-term follow-up data; robust comparative long-term follow-up data are lacking. Conversely, monopolar TURP has stood the test of time as it has proven long-lasting improvement of patient symptoms with a low reoperation rate. Therefore, monopolar TURP will continue to remain the reference standard surgical intervention procedure for men with voiding LUTS due to BPH for the foreseeable future.

References

1. Cornu JN, Ahyai S, Bachmann A, De La Rosette J, Gilling P, Gratzke C, et al. A systematic review and meta-analysis of functional outcomes and complications following transurethral procedures for lower urinary tract symptoms resulting from benign prostatic obstruction: an update. *Eur Urol* 2015:1066–96.
2. Hinz S, Rais-Bahrami S, Kempkensteffen C, Weiske WH, Schrader M, Magheli A. Fertility rates following vasectomy reversal: importance of age of the female partner. *Urol Int* 2008;**81**(4):416–20.
3. Ahyai SA, Gilling P, Kaplan SA, Kuntz RM, Madersbacher S, Montorsi F, et al. Meta-analysis of functional outcomes and complications following transurethral procedures for lower urinary tract symptoms resulting from benign prostatic enlargement. *Eur Urol* 2010:384–97.
4. Rieken M, Bachmann A, Shariat SF. Long-term follow-up data more than 5 years after surgical management of benign prostate obstruction: who stands the test of time?. [Internet] *Curr Opin Urol* 2015:22–7. Available from: http://journals.lww.com/co-urology/pages/default.aspx%5Cnhttp://ovidsp.ovid.com/ovidweb.cgi?T=JS&PAGE=reference&D=emed13&NEWS=N&AN=2015503412.
5. Smith RD, Patel A. Transurethral resection of the prostate revisited and updated. *Curr Opin Urol* 2011;**21**(1):36–41.
6. Rassweiler J, Teber D, Kuntz R, Hofmann R. Complications of transurethral resection of the prostate (TURP)-Incidence, management, and prevention. *Eur Urol* 2006:969–80.
7. Issa MM. Technological advances in transurethral resection of the prostate: bipolar versus monopolar TURP. [Internet] *J Endourol* 2008;**22**(8):1587–95. Available from: http://www.ncbi.nlm.nih.gov/pubmed/18721041.
8. Neal DE. The morbidity of transurethral resection of the prostate. *Curr Opin Urol* 1996;**6**(3):147–50.
9. Nakahira J, Sawai T, Fujiwara A, Minami T. Transurethral resection syndrome in elderly patients: a retrospective observational study. [Internet] *BMC Anesthesiol* 2014;**14**(1):30. Available from: http://www.pubmedcentral.nih.gov/articlerender.fcgi?artid=4004457&tool=pmcentrez&rendertype=abstract.
10. Walker KM. *British medical transurethral resection of the prostate a review of fourteen years' work.* 1937.
11. Ray EH. Transurethral resection of the prostate. [Internet] *J Urol* 1953;**69**(3):29–36. Available from: http://www.ncbi.nlm.nih.gov/pubmed/16385196.
12. Wu AK, Blaschko SD, Garcia M, McAninch JW, Aaronson DS. Safer urethral catheters: how study of catheter balloon pressure and force can guide design. *BJU Int* 2012;**109**:1110–4.

13. Bultitude MF. Campbell-walsh urology tenth edition. [Internet] *BJU Int* 2012;**109**:E10. Available from: http://doi.wiley.com/10.1111/j.1464-410X.2011.10907.x.
14. Herr HW. Max Nitze, the cystoscope and urology. *J Urol* 2006;**176**(4):1313–6.
15. Mayer EK, Kroeze SGC, Chopra S, Bottle A, Patel A. Examining the "gold standard": a comparative critical analysis of three consecutive decades of monopolar transurethral resection of the prostate (TURP) outcomes. *BJU Int* 2012;**110**(11):1595–601.
16. Rees J, Bultitude M, Challacombe B. The management of lower urinary tract symptoms in men. [Internet] *BMJ* June 24 , 2015;**348**:g3861. Available from: http://www.ncbi.nlm.nih.gov/pubmed/24963071%5Cnhttp://www.bmj.com/cgi/doi/10.1136/bmj.g3861.
17. McVary KT, Roehrborn CG, Alvins AL, Barry MJ, Bruskewitz RC, Donnell RF, et al. American urological association guideline: management of benign prostatic hyperplasia (BPH). [Internet] *Am Urol Assoc* 2010:496. Available from: https://www.auanet.org/common/pdf/education/clinical-guidance/Benign-Prostatic-Hyperplasia.pdf.
18. Gravas S, Bach T, Bachmann A, Drake M, Gacci M, Gratzke C, et al. Guidelines on the management of non-neurogenic male lower urinary tract symptoms (LUTS), incl. benign prostatic obstruction (BPO). [Internet] *Eur Assoc Guidel* 2015:1–70. Available from: https://uroweb.org/wp-content/uploads/EAU-Guidelines-Non-Neurogenic-Male-LUTS-Guidelines-2015-v2.pdf.
19. Mcvary KT, Roehrborn CG, Avins AL, Barry MJ, Bruskewitz RC, Donnell RF, et al. Update on AUA guideline on the management of benign prostatic hyperplasia. *J Urol* 2011;**185**:1793–803.
20. Bruskewitz RC, Iversen P, Madsen PO. Value of postvoid residual urine determination in evaluation of prostatism. *Urology* 1982;**20**(6):602–4.
21. Mebust WK, Holtgrewe HL, Cockett AT, Peters PC. Transurethral prostatectomy: immediate and postoperative complications. A cooperative study of 13 participating institutions evaluating 3,885 patients. [Internet] *J Urol* 1989;**141**(2):243–7. Available from: http://www.ncbi.nlm.nih.gov/pubmed/2643719.
22. Taylor BL, Jaffe WI. Electrosurgical transurethral resection of the prostate and transurethral incision of the prostate (monopolar techniques). *Can J Urol* October 2015;(22):24–9.
23. Raj MD, McDonald C, Brooks AJ, Drummond M, Lau HM, Patel MI, et al. Stopping anticoagulation before TURP does not appear to increase perioperative cardiovascular complications. *Urology* 2011;**78**(6):1380–4.
24. Davis NF, McGuire BB, Flood HD. Perioperative management of antithrombotic agents in urological surgery. [Internet]. British Association of Urological Surgeons *Br J Med Surg Urol* 2011;**4**(5):187–96. Available from: https://doi.org/10.1016/j.bjmsu.2010.08.004.
25. Davis NF, Fanning DM, McGuire BB, Carroll GT, Flood HD. Perioperative management of chronic anticoagulation therapy in urological patients: a cross-sectional survey of practice. *Ir J Med Sci* 2011;**180**(4):823–8.
26. Emberton M, Neal DE, Black N, Harrison M, Fordham M, McBrien MP, et al. The national prostatectomy audit: the clinical management of patients during hospital admission. *Br J Urol* 1995;**75**(3):301–16.
27. Porter M, McCormick B. Anaesthesia for transurethral resection of the prostate (TURP). *Update Anaesth* 2003:21–6.
28. Bootsma AMJ, Laguna Pes MP, Geerlings SE, Goossens A. Antibiotic prophylaxis in urologic procedures: a systematic review. *Eur Urol* 2008:1270–86.
29. Babayan RK. Re: assessment of antibiotic prophylaxis prescribing patterns for TURP: a need for Canadian guidelines? *J Urol* 2014:371.
30. Hahn RG, Fagerström T, Tammela TLJ, Van Vierssen Trip O, Beisland HO, Duggan A, et al. Blood loss and postoperative complications associated with transurethral resection of the prostate after pretreatment with dutasteride. *BJU Int* 2007;**99**(3):587–94.

31. Kavanagh LE, Jack GS, Lawrentschuk N. Prevention and management of TURP-related hemorrhage. [Internet] *Nat Rev Urol* 2011;**8**(9):504–14. Available from: http://www.ncbi.nlm.nih.gov/pubmed/21844906.
32. Sudhindra TV, Joseph A, Haray PN, Hacking BC. Are surgeons aware of the dangers of diathermy? *Ann R Coll Surg Engl* 2000;**82**(6 Suppl.):189–90.
33. Holman CDJ, Wisniewski ZS, Semmens JB, Rouse IL, Bass AJ. Mortality and prostate cancer risk in 19,598 men after surgery for benign prostatic hyperplasia. *BJU Int* 1999;**84**(1):37–42.
34. Varkarakis J, Bartsch G, Horninger W. Long-term morbidity and mortality of transurethral prostatectomy: a 10-year follow-up. *Prostate* 2004;**58**(3):248–51.
35. Jensen V. The TURP syndrome. *Can J Anaesth* 1991;**38**(1):90–7.
36. Vijayan S. TURP syndrome. *Trends Anaesth Crit Care* 2011:46–50.
37. Tasc AI, Ilbey YO, Tugcu V, Cicekler O, Cevik C, Zoroglu F. Transurethral resection of the prostate with monopolar resectoscope: single-surgeon experience and long-term results of after 3589 procedures. *Urology* 2011;**78**(5):1151–5.
38. Muntener M, Aellig S, Kuettel R, Gehrlach C, Sulser T, Strebel RT. Sexual function after transurethral resection of the prostate (TURP): results of an independent prospective multicentre assessment of outcome. *Eur Urol* 2007;**52**(2):510–6.
39. Fagerström T, Nyman CR, Hahn RG. Complications and clinical outcome 18 months after bipolar and monopolar transurethral resection of the prostate. [Internet] *J Endourol* 2011;**25**(6):1043–9. Available from: http://www.ncbi.nlm.nih.gov/pubmed/21568691.
40. Flanigan RC, Reda DJ, Wasson JH, Anderson RJ, Abdellatif M, Bruskewitz RC, et al. 5-Year outcome of surgical resection and watchful waiting for men with moderately symptomatic benign prostatic hyperplasia: a Department of Veterans Affairs Cooperative Study. *J Urol* 1998;**160**(1):12–7.
41. Meyhoff HH, Nordling J. Long term results of transurethral and transvesical prostatectomy. A randomized study. [Internet] *Scand J Urol Nephrol* 1986:27–33. Available from: http://onlinelibrary.wiley.com/o/cochrane/clcentral/articles/682/CN-00042682/frame.html.
42. Mamoulakis C, Schulze M, Skolarikos A, Alivizatos G, Scarpa RM, Rassweiler JJ, et al. Midterm results from an international multicentre randomised controlled trial comparing bipolar with monopolar transurethral resection of the prostate. *Eur Urol* 2013;**63**(4):667–76.
43. Akman T, Binbay M, Tekinarslan E, Tepeler A, Akcay M, Ozgor F, et al. Effects of bipolar and monopolar transurethral resection of the prostate on urinary and erectile function: a prospective randomized comparative study. *BJU Int* 2013;**111**(1):129–36.
44. Omar MI, Lam TB, Alexander CE, Graham J, Mamoulakis C, Imamura M, et al. Systematic review and meta-analysis of the clinical effectiveness of bipolar compared with monopolar transurethral resection of the prostate (TURP). *BJU Int* 2014;**113**(1):24–35.
45. Rieken M, Bachmann A, Shariat SF. Long-term follow-up data more than 5 years after surgical management of benign prostate obstruction: who stands the test of time? *Curr Opin Urol* 2016;**26**(1):22–7.
46. Issa MM, Young MR, Bullock AR, Bouet R, Petros JA. Dilutional hyponatremia of TURP syndrome: a historical event in the 21st century. *Urology* 2004;**64**(2):298–301.

Bipolar Plasma Kinetic Vaporization

8

Stephanie Tsai[1], *Dominique Thomas*[1], *Bilal Chughtai*[1], *Steven Kaplan*[2]
[1]Weill Cornell Medicine, New York, NY, United States; [2]Icahn School of Medicine, New York, NY, United States

Introduction

Bipolar plasma kinetic vaporization of the prostate (BPKVP) is an endoscopic transurethral ablation technique used in the treatment of lower urinary tract symptoms (LUTS) secondary to benign prostatic hyperplasia (BPH). The BPKVP system makes use of electrical principles explored in earlier forms of transurethral high-frequency surgery. Also known as the "button" procedure, BPKVP is an alternative to traditional monopolar transurethral resection of the prostate (mTURP). This procedure utilizes an axipolar electrode in the shape of a mushroom, or "button," and electroconductive solution to apply low-temperature radiofrequency plasma energy to vaporize prostate tissue. The following chapter will navigate through the development, outcomes, and complications of the bipolar plasma kinetic vaporization in the treatment of BPH.

Monopolar Electrovaporization Technology

The application of energy in the form of an electric current leads to a rapid rise in tissue temperature, causing boiling of intracellular water and subsequent cellular disruption and destruction. The vaporization process is localized to the tissue surrounding the electrode. One major disadvantage of vaporizing techniques is that they do not provide tissue for histological examination.[1]

Prior to the introduction of bipolar cautery techniques, transurethral electrovaporization of the prostate (TEVP) was developed as a similar early alternative to mTURP.[2] This approach utilizes a monopolar roller ball electrode to apply high-voltage cutting currents to remove the prostate tissue. The grooved roller ball electrode consists of a leading edge that vaporizes tissue and a trailing edge that coagulates and seals vessels in vaporized tissue. As a result, electrovaporization allows for the simultaneous vaporization, desiccation, and coagulation of prostatic tissue.

Early studies of TEVP demonstrated the procedure's safety and efficacy for the treatment of LUTS associated with BPH. Kaplan et al. conducted an 18-month follow-up of 114 patients with LUTS who underwent TEVP.[3] At 18 months, patients experienced a decrease in American Urological Association (AUA) symptom score (−11.3, $P<.01$) and an increase in peak urinary flow rate (Qmax+9.6 mL/s, $P<.001$). There was no significant change in postvoid residual (PVR), measured at 0 and 12 months (64.5 vs. 33.6 mL, $P=.07$). The mean catheterization time was 10.4 h, mean hospital stay was

A Comprehensive Guide to the Prostate. https://doi.org/10.1016/B978-0-12-811464-3.00008-8

0.9 days, and mean operative time was 47.8 min. No patients experienced postoperative complications of transurethral resection (TUR) syndrome or significant blood loss requiring transfusion.

In a prospective randomized trial of 64 patients with moderate to severe LUTS, TEVP was found to have similar efficacy to mTURP.[4] At 12 months, the TEVP group had greater reduction in AUA symptom score, compared with the mTURP group (−12.8 vs. −12.2, $P<.02$). However, the mTURP group experienced greater increase in Qmax (+11.3 v. +9.7 mL/s, $P<.03$). There was no statistically significant difference in change in PVR between the two groups. The TEVP group had longer operative time (47.6 vs. 34.6 min, $P<.003$) but had shorter catheterization time (12.9 vs. 67.4 h, $P<.01$) and shorter total hospitalization time (1.3 vs. 2.6 days, $P<.03$). The mTURP group had significantly larger decreases in sodium levels (−3.9 vs. −1.4 mEq/L, $P<.03$). There were no major complications in the TEVP group. In contrast, in the mTURP group, one patient required transfusion, and one patient developed TUR syndrome.

Despite the similar efficacy of mTURP and TEVP in relieving LUTS due to BPH, TEVP has largely been abandoned due to the excessive extent of coagulation. In TEVP, the disproportionate extent of coagulation (up to 10 mm) has been shown to result in higher incidence of irritative micturition symptoms and stress incontinence.[5,6] In contrast, bipolar approaches to prostate tissue vaporization result in a more limited energy delivery.

The Bipolar Plasma Kinetic Vaporization Approach

Bipolar plasma kinetic vaporization utilizes two isolated electrodes (active and return) that form a complete circuit with the tissue. This axipolar electrode, shaped as a mushroom or "button," operates at 160 Ω in a radiofrequency range of 320–350 Hz and a voltage range of 254–350 V.[7] Saline irrigant serves as the conductive medium and is converted by the radiofrequency energy into a plasma field, producing vaporization of tissue on contact while achieving hemostasis.[8] As a result, the current is transmitted locally and flows only through the tissue between the poles of the electrodes, allowing for safer and more precise energy delivery. Like TEVP, the BPKVP approach does not allow for preservation of tissue for histological examination.

In monopolar devices, the active electrode and return electrode (skin pad) are separated by a greater distance, thus requiring higher voltages during resection. The incorporation of both the active and return electrodes within bipolar devices limits tissue impedance and therefore requires lower sustained voltages during resection.[9] Because resection can occur at lower voltages and temperatures in bipolar systems, BPKVP has the theoretical advantages of eliminated risk of inadvertent skin burns, reduced collateral tissue injury, reduced risk of nerve injury, and decreased interference with cardiac pacemakers. In mTURP, a nonconductive solution for continuous irrigation is used, whereas BPKVP utilizes a normal saline irrigant. As a result, the BPKVP approach is also hypothesized to reduce the risk of development of hyponatremia associated with TUR syndrome.

Results

Botto et al. reported preliminary clinical results using the BPKVP technique in 42 patients, with follow-up of 3 months.[10] At 3 months, patients had a reduction in International Prostate Symptom Score (IPSS, from 16 to 9), and an increase in Qmax (7.9–19.7 mL/s). Postoperative complications of TUR syndrome and significant blood loss requiring transfusion were not seen in any patients. Similarly, in a study of 30 patients who underwent BPKVP, there was a reduction in IPSS (−12.7, $P<.01$) and an increase in Qmax (+11.5, $P<.01$) at 6 months.[11] No major complications occurred intra- or postoperatively, and no blood transfusions were required. However, these studies had small nonrandomized cohorts with limited follow-up.

Larger randomized trials comparing the results of BPKVP and mTURP have generally found reduced operative, catheterization, and hospitalization times for BPKVP. In a prospective study of 75 patients with LUTS secondary to BPH, patients who underwent BPKVP were found to have shorter operation times (40.5 vs. 55 min, $P<.001$), shorter postoperative irrigation periods (6 vs. 20 h, $P<.001$), and earlier catheter-removal times (35 vs. 68 h, $P<.001$) than those who underwent mTURP.[7] Geavlete et al. conducted a prospective, randomized study comparing BPKVP with mTURP in 155 patients, with a 6-month follow-up period.[12] Patients in the BPKVP group experienced shorter operative duration (35.1 vs. 50.4 min, $P=.002$), catheterization period (23.8 vs. 71.2 h, $P=.002$), and hospital stay (47.6 vs. 93.1 h, $P=.018$). In a further randomized trial by the same group, BPKVP was compared with mTURP and bipolar transurethral resection of the prostate (bTURP), with a follow-up period of 18 months.[13] This larger study, which included 170 patients in each study arm, found that the BPKVP had significantly shorter operation time compared to the mTURP and bTURP groups (39.7 vs. 52.1 and 55.6 min, respectively, $P=.0001$). In addition, catheterization time and hospital stay were significantly reduced for the BPKVP group. The shorter operative times associated with BPKVP have been attributed to rapid hemostasis, maintenance of visibility due to decreased bleeding, and removal of debris from subjacent tissue.

BPKVP has also been shown to have better efficacy, as measured by IPSS and Qmax, in short- and medium-term follow-up periods as compared with mTURP. For example, Karaman et al. found that patients in the BPKVP group experienced greater improvement IPSS as compared with the mTURP group at 3 months (−14 vs. −10, $P<.001$).[7] In their 18-month follow-up of patients in the BPKVP and mTURP groups, Geavlete et al. found that, despite having similar preoperative parameters, BPKVP had a significantly greater decrease in IPSS (−19.3 vs. −15.9, $P<.0001$) and a significantly greater increase in Qmax (+17.1 vs. +13.8 mL/s, $P<.0001$).[13] Prostate volume and prostate-specific antigen (PSA) values were comparable.

For longer follow-up periods, there is limited evidence to support better efficacy of BPKVP over mTURP. Muslumanoglu et al. conducted a randomized trial of 101 patients, with follow-up at 12, 60, and 100 months, and found that BPKVP had similar long-term results to mTURP.[14] At 12 months, the mean improvement rate from baseline Qmax was significantly greater in the BPKVP group compared to the mTURP group (120.5% vs. 103.6%, $P<.05$). However, at 100 months, mean Qmax levels in both groups had decreased to similar levels (15.9 vs. 15.9 mL/s, $P=.34$). Reoperation was performed in six patients in the BPKVP group and four patients in the mTURP group at the end of 100 months. However, approximately 30% of patients were lost to follow-up at 100 months.[14] There remains limited research on the long-term efficacy and reoperation rates of BPKVP compared to mTURP.

Complications

In studies comparing BPKVP to mTURP, BPKVP is associated with fewer postoperative complications. In a prospective study with 12 months follow-up, Karaman et al. reported that two patients in the mTURP group experienced blood loss requiring transfusion, compared with no patients in the BPKVP group.[7] Both groups had similar incidences of urethral strictures, bladder–neck strictures, and erectile dysfunction.[7] Geavlete et al. found that, compared to BPKVP, mTURP was associated with a greater number of capsular perforations (7 of 80 for mTURP vs. 1 of 75 for BPKVP, $P=.037$) and cases of significant intraoperative bleeding (11 of 80 for mTURP vs. 1 of 75 for BPKVP, $P=.013$).[12] The mTURP group had a mean hemoglobin decrease of 1.5 g/dL, compared to a 0.6 g/dL decrease in the BPKVP group ($P=.002$). In a larger study with 18 months follow-up, Geavlete et al. further found that, compared to the mTURP group, the BPKVP group had significantly decreased rates of postoperative hematuria, blood transfusion, clot retention, recatheterization, and rehospitalization for secondary hemorrhage.[13] In addition, the BPKVP group had significantly reduced rates of bladder–neck sclerosis (0.6% vs. 4.1%, $P=.047$) at 18 months. The rates of early irritative symptoms, dysuria, urgency, frequency, nocturia, urinary tract infections, urethral strictures, and urinary incontinence were similar in both groups. In studies with longer-term follow-up periods (up to 100 months), BPKVP and mTURP were found to have no difference in urethral stricture rates.[14]

The use of normal saline in BPKVP has been shown to result in decreased risk of TUR syndrome as compared with mTURP, which uses nonconductive irrigant. In a study of 152 patients who underwent BPKVP or mTURP, no patients in the BPKVP group developed TUR syndrome or had a significant decrease in hemoglobin.[15] In contrast, 2 of 100 patients in the mTURP group developed TUR syndrome in the form of mental confusion and irritability with serum sodium <125 mEq/L. Elsakka et al. in a prospective randomized trial found that there was a significant perioperative drop in serum sodium (from 137.5 to 129.4 mEq/L, $P<.001$) in the mTURP group, and TUR syndrome developed only in the mTURP group.[16] Overall, BPKVP appears to be associated with fewer perioperative and postoperative complications compared to mTURP and specifically avoids TUR syndrome.

Conclusion

BPKVP represents a viable alternative in the treatment of LUTS secondary to BPH. Although mTURP has long been considered the gold standard of therapeutic approaches to BPH, it is still associated with significant peri- and postoperative complications and morbidity. With the emergence of alternative techniques, including BPKVP, mTURP has declined in popularity since 2000.[17] BPKVP, with its shorter operative, catheterization, and hospitalization time, may be an attractive alternative to the traditional mTURP approach. In addition, trials comparing BPKVP to mTURP have demonstrated similar or better outcomes in terms of IPSS, Qmax, and PSA levels in short- to medium-term follow-up periods. BPKVP is also associated with fewer perioperative and postoperative complications. While more data are needed on the long-term efficacy of BPKVP compared to mTURP, existing research demonstrates that BPKVP serves as a safe and effective alternative.

References

1. Kuntz RM. Laser treatment of benign prostatic hyperplasia. *World J Urol* 2007;**25**(3):241–7.
2. Kaplan SA, Te AE. A comparative study of transurethral resection of the prostate using a modified electro-vaporizing loop and transurethral laser vaporization of the prostate. *J Urol* 1995;**154**(5):1785–90.
3. Kaplan SA, Santarosa RP, Te AE. Transurethral electrovaporization of the prostate: one-year experience. *Urology* 1996;**48**(6):876–81.
4. Kaplan SA, Laor E, Fatal M, Te AE. Transurethral resection of the prostate versus transurethral electrovaporization of the prostate: a blinded, prospective comparative study with 1-year followup. *J Urol* 1998;**159**(2):454–8.
5. Gallucci M, Puppo P, Perachino M, et al. Transurethral electrovaporization of the prostate vs. transurethral resection. Results of a multicentric, randomized clinical study on 150 patients. *Eur Urol* 1998;**33**(4):359–64.
6. Reich O, Corvin S, Oberneder R, Sroka R, Muschter R, Hofstetter A. In vitro comparison of transurethral vaporization of the prostate (TUVP), resection of the prostate (TURP), and vaporization-resection of the prostate (TUVRP). *Urol Res* 2002;**30**(1):15–20.
7. Karaman MI, Kaya C, Ozturk M, Gurdal M, Kirecci S, Pirincci N. Comparison of transurethral vaporization using plasmakinetic energy and transurethral resection of prostate: 1-year follow-up. *J Endourol* 2005;**19**(6):734–7.
8. Tefekli A, Muslumanoglu AY, Baykal M, Binbay M, Tas A, Altunrende F. A hybrid technique using bipolar energy in transurethral prostate surgery: a prospective, randomized comparison. *J Urol* 2005;**174**(4 Pt 1):1339–43.
9. Maddox M, Pareek G, Al Ekish S, et al. Histopathologic changes after bipolar resection of the prostate: depth of penetration of bipolar thermal injury. *J Endourol* 2012;**26**(10):1367–71.
10. Botto H, Lebret T, Barré P, Orsoni JL, Hervé JM, Lugagne PM. Electrovaporization of the prostate with the Gyrus device. *J Endourol* 2001;**15**(3):313–6.
11. Reich O, Schlenker B, Gratzke C, et al. Plasma vaporisation of the prostate: initial clinical results. *Eur Urol* 2010;**57**(4):693–7.
12. Geavlete B, Multescu R, Dragutescu M, Jecu M, Georgescu D, Geavlete P. Transurethral resection (TUR) in saline plasma vaporization of the prostate vs standard TUR of the prostate: 'the better choice' in benign prostatic hyperplasia? *BJU Int* 2010;**106**(11):1695–9.
13. Geavlete B, Georgescu D, Multescu R, Stanescu F, Jecu M, Geavlete P. Bipolar plasma vaporization vs monopolar and bipolar TURP-A prospective, randomized, long-term comparison. *Urology* 2011;**78**(4):930–5.

14. Muslumanoglu AY, Yuruk E, Binbay M, Akman T. Transurethral resection of prostate with plasmakinetic energy: 100 months results of a prospective randomized trial. *BJU Int* 2012;**110**(4):546–9.
15. Aboutaleb H. Efficacy of bipolar "button" plasma vaporization of the prostate for benign prostatic obstruction, compared to the standard technique. *Urol Ann* 2015;**7**(4):442–7.
16. Elsakka AM, Eltatawy HH, Almekaty KH, Ramadan AR, Gameel TA, Farahat Y. A prospective randomised controlled study comparing bipolar plasma vaporization of the prostate to monopolar transurethral resection of the prostate. *Arab J Urol* 2016;**14**(4):280–6.
17. Malaeb BS, Yu X, McBean AM, Elliott SP. National trends in surgical therapy for benign prostatic hyperplasia in the United States (2000-2008). *Urology* 2012;**79**(5):1111–6.

532 nm Laser Photoselective Vaporization of the Prostate

9

Benjamin E. Muller, Gina M. Badalato, Matthew P. Rutman
New York Presbyterian Hospital/Columbia University Medical Center, New York, NY, United States

Introduction

There is a long history of advancement in minimally invasive therapies for benign prostatic hyperplasia (BPH). The gold standard in the treatment of BPH has long been transurethral resection of the prostate (TURP), performed using an electrocautery loop under direct visualization with a cystoscope.[1] Over the past two decades, however, the use of lasers for photoselective vaporization of the prostate (PVP) has become more common in treatment of BPH, with concomitant advances in the efficacy and utility of this technology.[2] In this chapter, we review the history of this procedure and relevant technical details of the lasers used in PVP. We then discuss surgical technique, outcomes, including comparison to other modalities, and relative cost of the procedure.

History

The genesis of the laser (from LASER, light amplification by stimulated emission of radiation) can be traced to Einstein. However, the first true laser was not developed until 1960 by Theodore Maiman, building on the work of James Gordon and others. In its most basic form, a laser is a laser medium in a mirrored chamber that repeatedly reflects photons through the medium, allowing for the stimulated emission and subsequent amplification from which the name laser is derived. Notably, the light emitted by a laser is of a uniform wavelength, frequency, and energy (i.e., laser light is coherent).[2]

The use of lasers for medical therapy began shortly after their initial development. Charles Campbell, an ophthalmologist at Columbia University, treated the first human patient in November 1961, little more than a year after Maiman first announced his laser.[3] Since then, lasers have become nearly ubiquitous in medicine and are used in numerous capacities across many specialties. In urology, lasers have been used most notably for treatment of BPH and stone disease but have also been studied in ablation of cancers, treatment of strictures, and even for "welding" anastomoses.[2]

Lasers were first used for treatment of BPH in the late 1980s and early 1990s, and the neodymium-YAG (Nd:YAG) laser predominated during this early period. This laser has a wavelength of 1064 nm (outside the visible spectrum), and a penetration depth of approximately 1 cm. Used in a noncontact fashion (called visual laser ablation of

A Comprehensive Guide to the Prostate. https://doi.org/10.1016/B978-0-12-811464-3.00009-X

the prostate), it causes a zone of necrotic tissue that ultimately sloughs off, creating an enlarged cavity and reducing symptoms of BPH. However, this sloughing process and the associated edema can cause significant urinary retention and lower urinary tract symptoms and may result in prolonged catheterization for a period of months. The Nd:YAG laser can also be used in direct contact with prostatic tissue, which directly vaporizes the tissue and eliminates the symptoms of sloughing. However, this process is slow, making it unsuitable for larger prostate glands (over 40 g). Moreover, the hemostasis achieved with direct contact is not as good as with the noncontact mode.[2,4]

Development of the potassium titanyl phosphate (KTP) laser led to significant changes in the technique and outcomes of laser treatment for BPH. KTP is not an entirely new laser, but rather the addition of a KTP crystal to an Nd:YAG laser, which results in a doubling of laser frequency and halving of wavelength to 532 nm. This brings the emitted light into the visible spectrum, hence the moniker "GreenLight."[5] The shorter wavelength provides several advantages. 532 nm light is highly absorbed by hemoglobin, so that in vascular tissue absorption is high and tissue temperature quickly surpasses the boiling point, leading to vaporization. In the area surrounding vaporization, tissue temperatures are sufficient for coagulation but not vaporization, providing a zone of hemostasis. Moreover, the KTP laser is effectively unabsorbed by aqueous media (the term photoselective vaporization comes from the selective absorption by hemoglobin and not the transfer medium) meaning that a noncontact visual technique can be used for vaporization. As a result, however, the laser can travel great distances in water or saline and a side-firing laser must be used to maintain careful control of the beam.[4]

The KTP laser itself has undergone a number of changes since it was introduced. The initial version of the laser was relatively underpowered at 40 Watts (W). Subsequent iterations increased power, to 60 and then 80 W. The KTP crystal itself was then replaced by lithium triborate (LBO), which creates the same 532 nm beam but with much greater efficiency, allowing for lasers of increased power—currently up to 180 W. LBO lasers also increased the working distance range (0.5–3 mm), compared to 0.5 mm with the original KTP, permitting better visualization of the ablative process.[2]

As the laser used in PVP has improved, so too has the laser fiber itself. The heat produced by the laser affects not only the adenomatous prostate but also the fiber as well. Carbonization and melting of the fiber tip can negatively impact the power and handling characteristics of the laser, prolonging PVP surgeries. Hermanns et al. examined the power characteristics of the 120 W LBO laser (high performance system, HPS) and found that power output dipped by more than 50% from the beginning to the end of the case secondary to fiber damage. This effect was worsened by tissue contact and especially prostate stones, which can cause complete fiber destruction. This damage to the fiber can be clinically significant if it prevents full vaporization of the prostate; 33% of patients in this study reported voiding symptoms, with 1 of 35 requiring reoperation due to persistent adenomatous tissue.[6] However, with the third and latest generation 180 W LBO laser (XPS), a new fiber has been introduced that uses a constant saline flow to cool the fiber tip (the MoXy fiber). In addition, it has FiberLife, which provides temperature sensing feedback to preserve fiber integrity by activating standby mode when increased temperatures are sensed. Results from both in vivo and in vitro testing of this fiber by Brunken et al. suggest that it does not suffer from the same durability issues as in previous generations, with no loss in maximum power output. This held true even after using the full 650kJ life span of the fiber during PVP.[7]

Surgical Considerations

Surgical Technique

PVP may be performed under general anesthesia or spinal anesthetic. The patient should be placed in dorsal lithotomy and prepped and draped in the same fashion as for a TURP. Preoperative antibiotics should be administered according to American Urological Association guidelines and/or prior urine culture sensitivities. Prior to beginning surgery, appropriate safety precautions must be taken. In particular, all personnel in the room during the procedure should wear appropriate eye protection that filters the 532nm light produced by the laser. A break in the laser fiber could cause unintended escape of high energy light, which can cause serious ocular damage.

To perform a PVP, we use a 23 Fr continuous flow cystoscope with a GreenLight filter for protection of the camera's charge-coupled device sensor. For larger prostate glands, actively anticoagulated patients, and those with concomitant bladder stones, we often use a 26 French resectoscope with a laser bridge. This improves visualization by increasing irrigation flow. We typically use a 30-degree lens for the cystoscope, as this allows for the best visualization of the prostatic urethra and ureteral orifices. Some manufacturers (e.g., Storz) make a 30-degree lens with a built-in filter; when this is used, there is no need to place the additional external filter. Two irrigation systems are needed: one for the scope, usually consisting of two 3L bags of normal saline connected by a Y-adapter, and one for the integrated cooling of the MoXy fiber, typically a 1L bag of normal saline. The cystoscope should always be introduced with a visual obturator, which reduces urethral trauma and unnecessary bleeding. Atraumatic passage of the scope is critical to optimize visibility.

The laser fiber has two important markings at the tip that should be visible at all times during the procedure. The blue triangle on the fiber cap is directly opposite to the exit site of the 532nm beam. Maintaining visibility of this triangle throughout the procedure ensures that the cystoscope is not damaged by the beam. The red octagon 180degrees around the fiber from the blue triangle indicates the aiming beam, which is a low-powered red laser aligned with the main beam. The laser should not be fired unless the aiming beam can clearly be seen on prostatic tissue.

Unlike a TURP, in which the electrocautery loop is pulled toward a stabilized cystoscope as tissue is excised, in PVP the fiber tip and cystoscope must move together. This both allows for better visualization of the vaporization process and prevents accidental retraction of the laser into the cystoscope, where it can cause significant damage. Moreover, in a TURP the typical motion of the instrument is axial, in and out of the cystoscope. In PVP, by contrast, the motion is rotational and is often described as sweeping. In fact, the size of the sweep is very important—it must be large enough to make progress, but small enough to allow for sufficient vaporization in each pass. An analysis of vaporization rate by sweeping angle performed by Ko et al. determined that the ideal sweeping angle is between 15 and 30 degrees at both 120 and 180 W. This resulted in the highest vaporization rate, measured as mL of tissue vaporized per minute in their model using porcine kidneys.[8]

Another important difference from the standard TURP procedure is working distance. Whereas the electrocautery loop is pulled directly through tissue in a TURP, the laser fiber must be maintained 0.5–3 mm distant from the tissue for maximal efficiency. Some authors have suggested using the fiber cap width (1.8 mm) as a visual estimator for appropriate working distance.[9] Ko et al. found that increasing sweep angle above 30 degrees results in an increased distance to tissue, which increases the area of the tissue exposed to coagulative temperatures and decreases the exposure to vaporizing temperatures.[8] With appropriate sweep speed and tissue distance, large bubbles will be produced as tissue is vaporized. Smaller bubbles are indicative of less efficient vaporization and likely coagulation and suggest that the working distance is too large.

At the beginning of the procedure, the laser is set to 80 W to assess the effect of a low power setting on prostatic tissue. The power is then increased in 10 or 20 W increments until the desired effect is achieved. We typically titrate up to 180 W for the bulk of the procedure. PVP begins with the formation of a working space at the bladder neck, which is particularly important in patients who have large lateral lobes. This prevents tissue contact with the laser fiber and is imperative for protection of the fiber and improves visualization while allowing for good flow of irrigant. Once a working space is established, some practitioners define the limits of the ablation using a low power setting on the laser, typically at the 5 and 7 o'clock positions at the bladder neck and just proximal to the verumontanum. It is critical that the laser not be used distal to the verumontanum to avoid injuring the urethral sphincter.[9]

With these limits defined, the median lobe and bladder neck are vaporized using the sweeping motion described previously. Channels are created at 5 and 7 o'clock from the bladder neck to the verumontanum to delineate a target depth for the procedure. The incisions are carried down in a transurethral incision of the prostate like fashion until the surgical capsule is identified. For large median lobes, an additional 6 o'clock incision is made and again carried down to the capsule. The median lobe is then debulked working between 5 and 6 and 6 and 7 o'clock, respectively. The lateral lobes are approached next, and the laser is applied in symmetrical fashion. We approach the lateral lobes by starting anteriorly. For the left lateral lobe, the initial sweep is started at 1 o'clock and eventually continued down to the prior 5 o'clock groove from the median lobe/bladder neck. The right lateral lobe is started at 11 o'clock and brought down to 7 o'clock in a similar fashion. The goal should be to have a smooth surface throughout the procedure, targeting tissue peaks as they appear and vaporizing evenly rather than focusing in only one area. A smooth surface allows for better visualization and better control of bleeding if it occurs as compared to a more irregular surface. When performing the anterior ablation, it is crucial to extend the fiber far past the "hood" of the cystoscope to prevent damage. As long as the fiber is safely extended, the anterior vaporization can be accomplished without scope rotation. Careful attention must be paid to use less energy in cases where there is a thinner anterior lobe.

Finally, the apex is ablated, again taking care not to vaporize tissue beyond the verumontanum. Many practitioners reduce power to 80–120W to further reduce the risk of serious injury to the sphincter.[9] The end goal is to achieve a TURP-like defect down to the surgical capsule.

Although bleeding is usually much less significant in PVP than in TURP, it is a known complication of the procedure. Bleeding is most likely at the median lobe, bladder neck, and verumontanum. If bleeding is not severe, continuing the procedure proximally will often allow the scope to tamponade the distal bleeder. If hemostasis is not achieved, small vessels can be cauterized using the coagulation footswitch on the HPS or XPS lasers. For bleeding from larger vessels, which may significantly impair visibility, circumferential painting of the area with the laser is often successful. If bleeding cannot be stopped using this method, the Bugbee, rolling ball electrocautery, or bipolar cautery loop can be used after converting to irrigation with sterile water. A short period of tamponade using a catheter with a filled 30mL balloon on traction may also be used to attain hemostasis.[10]

Incision and Enucleation

The LBO laser used for PVP is just one of many lasers used in urology. As discussed in another chapter, a holmium laser can be used for enucleation or resection, rather than vaporization, of the prostate (holmium laser enucleation/resection of the prostate, HoLEP/HoLRP). This technique is typically used for larger prostates where vaporization or serial resection as in TURP would result in a very long procedure. Instead of gradually removing prostatic tissue from the lumen toward the capsule, HoLEP incises to the capsule and removes an entire lobe, which is then morcellated in the bladder prior to removal. Results from HoLEP have been good, though it is generally less available than PVP and has a much steeper learning curve.[11,12]

As the lasers used in PVP have become more powerful, they have been used to treat larger prostates, a natural result of which has been the exploration of incisional techniques using LBO lasers. Dr. Gomez-Sancha pioneered this technique and describes a spectrum from vaporization to en bloc enucleation, between which is a hybrid technique called vapoenucleation (PVEP). There are significant differences in technique between the enucleation and vaporization techniques. Most notably, enucleation may require the laser fiber to make contact with tissue, which is generally to be avoided in vaporization. The 2090 fiber, used in HPS systems, may be preferable in this case as it is more maneuverable and more resistant to contact damage than the liquid-cooled MoXy fiber used in XPS. In addition to changes in fiber handling, the resectoscope is used to aid in dissection, unlike in traditional PVP where both scope and laser should be kept off the tissue being treated.[11]

Results with vapoenucleation and vapor incision (VIT) have been promising. A small noninferiority trial with 103 patients compared HoLEP to PVEP with XPS and found that PVEP was noninferior in terms of symptom and quality of life scores at 12months. However, two important postoperative measures—urinary flow rate (Q_{max}) and prostate volume reduction—were significantly higher in HoLEP, suggesting that the improvements seen with PVEP may be less durable.[13] A retrospective, multicenter study of 956 patients used propensity score matching to compare PVP to VIT and found that VIT was associated with significantly more improvement in symptom scores, quality of life, and decreases in postvoid residual volume (PVR) and prostate-specific antigen (PSA) levels. Moreover, patients undergoing VIT had significantly fewer clinic visits in the early postoperative period (within 90 days of the procedure). PVP retained an important advantage in cost, though, as VIT required longer operative times and greater use of MoXy fibers: more than 15% of VIT cases used two or more MoXy fibers, while fewer than 5% of PVP cases did. This may be an argument in favor of greater use of the 2090 fiber for this procedure.[14] GreenLight incision and enucleation techniques have demonstrated encouraging results, but more data will be needed to make the case for PVEP or VIT over HoLEP. For now, the relative ease of PVP compared to incisional techniques remains an important advantage.

Postoperative Care

For most patients, PVP is an outpatient procedure. At the end of the procedure a 20 Fr two-way Foley catheter is placed. It is our practice to administer 10–20 mg of furosemide 10 min prior to completion of the surgery to act as a pharmacological continuous bladder irrigation (CBI). Following surgery, patients are transported to the recovery room and hydrated with IV and oral fluids. Patients with normal detrusor function and not in retention preoperatively can be given a trial of void, which is performed 6 h after the procedure. A bladder scan is performed prior to discharge. Most patients are discharged home with a catheter and are instructed on how to remove the catheter at home 24–48 h later (when the urine is clear). Patients may be prescribed a course of antibiotics according to surgeon preference; our practice is to limit the antibiotic to 24 h of dosing around the time of catheter removal. All patients are encouraged to avoid heavy lifting and Valsalva-like maneuvers for 10–14 days. In addition, we prescribe a stool softener to help with avoiding constipation. Anticoagulants should be restarted in discussion with the patient's primary doctor. If anticoagulation was not stopped prior to the procedure, the patient may be kept in the hospital for observation overnight to ensure there is no postoperative bleeding.[9] For these patients, we will elect to place a 22 Fr three-way catheter in case CBI is needed. In our experience, most of the time CBI is not needed and the patients are sent home with the CBI port plugged.

Learning Curve

Although one of the benefits of PVP is its relative user-friendliness, there is still a learning curve associated with the procedure. Even among experienced practitioners, those accustomed to TURP may need time to grow familiar with the two-handed style used in PVP for synchronous movement of the laser fiber and camera. Misrai et al. studied the learning curve for XPS performed by a urologist with no prior experience in PVP, over the first 200 patients. Two primary outcomes were considered: the ratio of lasing time to operative time, and the amount laser energy delivered to the prostate. The authors found a steady increase in the time ratio parameter; by the 125th patient, the ratio stabilized at 75%. The energy delivery target was achieved slightly sooner, with a 5kJ/mL goal reached by the 75th patient. The authors thus note that while these target numbers are certainly reachable, there is a clear learning curve for this procedure.[15]

Outcomes

Prior to human usage, the XPS laser was studied in dogs in comparison with the existing HPS laser. In this model system, Malek et al. reported significant improvements in key operative characteristics for XPS. XPS had a significantly better vaporization rate (mean 2.3mL/min vs. 1.3mL/min for HPS, $P=.03$) with a larger zone of coagulation (mean 2.0mm vs. 1.5mm, $P<.005$). Moreover, using the XPS laser the authors were able to create a significantly larger prostatic cavity, with a mean of 11.8mL versus 6.7mL for HPS ($P=.01$). As a result of the greater power and larger coagulation zone (CZ), these procedures were bloodless, whereas for HPS postoperative hematuria was noted. Despite the deeper CZ and larger cavity formation, healing was largely complete by 8weeks, with complete reepithelialization of the prostatic channel.[16]

Early studies in humans largely confirmed the safety and efficacy of XPS seen in dogs. A report from Bachmann et al. on 201 patients found substantial improvements in symptoms with few complications. International prostate symptom scores (IPSS) improved 10 points on average at 6months following the procedure and Q_{max} increased by a mean of 13mL/s. Similarly, PVR fell from 190 to 35 and PSA from 5.5 to 2.0at 6months. Serious complications (Clavien–Dindo Grade 3 or above) occurred in a small proportion of patients: 2.8% of patients for whom data were recorded experienced urinary retention, and 0.5% of patients (one) reported total inability to pass urine.[17]

The findings of these investigations have been borne out across multiple studies. A systematic review by Brunken et al. included 10 studies totaling 1640 patients. The authors concluded that XPS provides shorter operating times, improved fiber durability, and good volume reduction, without worsening side effects or adverse outcomes. However, they note that none of the 10 studies included long-term follow-up beyond 9months.[18] Several of the included studies, and others performed more recently, address specific questions including comparing XPS to other methods; these will be addressed more fully below and are summarized in Table 9.1.

XPS versus TURP

TURP is the gold standard for surgical management of BPH. In the early development of PVP, a number of randomized trials compared the 80 and/or 120W lasers to TURP, yet to date there has been only one randomized head-to-head trial of XPS versus TURP. The GOLIATH trial was a multicenter trial across nine countries in Europe that randomized 281 patients with symptomatic BPH, 139 to XPS, and 142 to TURP. The primary outcome of the trial was IPSS at 6months, and secondary outcomes included urine flow rate and complication-free rate. For these outcomes, there was no significant difference between the two methods, i.e., XPS was found to be noninferior to TURP.[19]

Table 9.1 **Comparison of XPS to Other Modalities**

Author	Bachmann	Ben-Zvi	Hueber	Rieken	
Year	2014	2013	2013	2013	
Comparison	TURP	HPS	HPS	HPS	80W KTP
Study type	RCT	Prospective cohort	Prospective cohort	Retrospective	
Patients (XPS)	281 (139)	200 (120)	1809 (622)	240 (80)	
Preoperative PV (mL)	2.4	−1.2	−6.65	0	11
Procedure time (min)	10.3	−36*	−27.45*	1	−4*
Active time (min)	10.7	−15*	−36.18*	0	N/A
Fiber usage	N/A	−0.5*	−1.18*	0	−0.1*
Hospitalization (days)	−1.3*	−1.2*	N/A	−1.6*	−1.4*
Foley removal (days)	−0.6*	−0.5*	N/A	0.1	0.3

Data given as (XPS mean value)—(comparison mean value). Significant values in bold. *, XPS advantage over comparison. *PV*, prostate volume; *RCT*, randomized controlled trail; *TURP*, transurethral resection of the prostate.

The results of the GOLIATH trial were updated at 12 months and again at 24 months. At 12 months, there remained no differences in complication rates or reintervention rates between the XPS and TURP groups. However, patients in the XPS group did self-report a higher rate of incontinence symptoms, as well as overactive bladder symptoms; while significant, these differences were too slight to be clinically meaningful.[20] At 2 years, XPS and TURP remained comparable in terms of IPSS, Q_{max}, and complication-free rate. Moreover, the reported slight advantage to TURP in incontinence and overactive bladder symptoms disappeared by 2 years. This randomized trial is the best available evidence that XPS is as safe and effective as TURP for the treatment of BPH.[21]

One possible difference between groups across the 2 years of the GOLIATH trial was seen in prostate cancer diagnosis. By the nature of PVP, no tissue is recovered for pathological analysis, while samples from TURP can be assessed for the presence of cancer. In the TURP group, five patients were diagnosed with prostate cancer at the time of the procedure, one within the first year, and one at 18 months. No patients in the XPS group were diagnosed during the first 2 years. This implies that there may be a surveillance advantage to TURP over XPS, although this study was not intended to discover differences in cancer diagnosis rates between groups.[21]

One other noteworthy comparison between XPS and TURP concerns fluid absorption during surgery. Historically, TURP was associated with the potential for a "TUR syndrome," which resulted from the use of a glycine solution for irrigation. Symptoms of the TUR syndrome can be serious and include nausea, vomiting, visual disturbances, and hypotension. With modern technique, bipolar cautery can be used for

TURP, and saline is used as the irrigating solution; TUR syndrome is thus no longer a major concern. However, fluid absorption during endoscopic procedures can still be of clinical importance, as in patients with heart failure or other comorbidities involving fluid balance.[22] Wettstein et al. assessed the extent of fluid absorption that occurs during PVP with XPS. The procedure was performed in the usual fashion, but an irrigation solution of isotonic saline with 1% ethanol was used in place of normal saline. Breath ethanol measurement was used to assess fluid absorption. Among the 54 patients in the study, 22 (41%) absorbed a detectable amount of irrigation fluid, in volumes ranging from 208 to 4579 mL. Nearly half of the patients who absorbed fluid took up more than 2 L; the median was just under 1 L. Absorbers had significant decreases in hemoglobin and hematocrit, as well as serum bicarbonate and venous pH compared to nonabsorbers, and increases in serum sodium and chloride. Five patients demonstrated symptoms possibly related to fluid absorption, including hypothermia, nausea, and somnolence with peripheral edema. These results demonstrate that while the TUR syndrome may be a relic of the past, fluid absorption during XPS remains an area of clinical concern.[23]

XPS With Anticoagulation

The ability to do a nearly bloodless procedure is one of the purported advantages of PVP over TURP, due to the regional coagulative effect of the laser. For that reason, PVP is a particularly appealing procedure in patients on anticoagulants, in whom TURP might be contraindicated. Studies in patients undergoing HPS have shown the procedure is safe for patients on anticoagulation: Chung et al. reported that PVP was safe in patients on aspirin, clopidogrel, warfarin, or a combination.[24] Other studies that included only patients taking one anticoagulant yielded similar results, both for aspirin and warfarin.[25,26]

Results with the 180W laser are consistent with these earlier studies. Lee et al. performed a retrospective analysis of 384 patients who underwent PVP with XPS, of whom 186 were on an anticoagulant—aspirin, clopidogrel, warfarin, or a combination (AC group). Compared to their counterparts (non-AC group), men in the AC group tended to be older (75 vs. 69.5 years old) with a higher American Society of Anesthesiologists (ASA) score, indicating more comorbid conditions (40.9% of patients had a score of three or above in the AC group, vs. 16.8% in the non-AC group). During the procedure, a significantly higher proportion of patients in the AC group required conversion to TURP (13.5% vs. 6.1%). This was particularly pronounced when patients were taking two or three anticoagulants: 25% of those patients were converted, as compared to 10% of patients on only one anticoagulant. Despite converting more procedures in the AC group, physicians experienced no difference in rates of difficult intraoperative bleeding between groups. Likewise, rates of postoperative hematuria, clot retention, and other bleeding complications were not significantly different between groups, and no patient in either group required blood transfusion. Though the AC group did stay a median of 12.5 h longer in the hospital ($P < .01$), given the equivalence of postoperative complication rates, this difference likely reflects only clinical caution.[27] XPS should be considered safe for use in patients on anticoagulation, though more head-to-head data are needed to determine if it is truly superior to TURP in such patients.

XPS in Large Prostates

Much as the coagulative effect of PVP represented an opportunity to treat anticoagulated patients for whom TURP was not the best option, the increased power of XPS has encouraged the use of PVP for patients with large prostates, in whom HPS was considered to be slow and ineffective relative to TURP. A number of investigators have assessed the use of XPS in large glands, which are variably defined to be greater than 80 or 100 mL. Altay et al. reported on operative and perioperative characteristics of patients with prostates greater than 80 mL treated with XPS. These results, including operative time, lasing time, and fiber usage are in line with the large prostate group described by Hueber et al. in their comparison with HPS discussed previously.[28,29]

In addition to their comparison of XPS and HPS, Hueber et al. describe outcomes in patients with large prostates. Nearly 1200 patients were included in a multicenter retrospective analysis, of whom 739 had a prostate of less than 80 mL (median 50 mL) and 387 had a prostate greater than 80 mL (median 108 mL). At 2 years follow-up, outcomes were generally comparable between groups; IPSS, quality of life score, and Q_{max} were not significantly different. PVR and PSA, however, remained significantly elevated in the large prostate group, though both groups saw substantial decreases in both values. Intraoperatively, the large prostate group had a significantly higher rate of conversion to TURP, with an odds ratio of 8.255 ($P<.01$). While reoperation rates were generally low in this study—just 1.2% at 2 years—patients who required retreatment had a mean prostate size of 120 mL, as compared to 70 mL in patients who did not need reoperation. Notably, the energy density in the group requiring reoperation was 2.1 kJ/mL, whereas it was 4.4 kJ/mL in patients not requiring retreatment ($P<.01$).[30]

Meskawi et al. report on a cohort of patients with glands greater than 100 mL (median 121 mL) with 4 years of follow-up. Increasing the threshold for large prostate to 100 mL leads to increases in operative time, energy use, and fiber use, as might be expected. Despite the increased prostate size, clinical outcomes remain good, with IPSS, Q_{max}, PVR, and PSA all significantly improved over 4 years. However, the authors note a potentially concerning degradation in these factors between years three and four, with increases in IPSS, PVR, and PSA, and a decrease in Q_{max}. Though these changes were not statistically significant, the trend is worrisome. In this group, as in the previously discussed group, patients requiring retreatment had lower energy density during the initial procedure (2.4 vs. 3.4 kJ/mL, $P=.02$) and a less durable PSA response (26% vs. 51% reduction at 12 months, $P=.02$).[31]

Some have suggested treating even larger glands by changing the technique to vapoenucleation or VIT. Stone et al. performed a prospective cohort study in 70 patients with glands greater than 150 mL using vapoenucleation. An increase in operative time, energy usage, and fiber usage was again seen, but once again, median outcomes significantly improved across categories (IPSS, quality of life, Q_{max}, PVR, PSA) at 1 year follow-up, with most remaining significant at 2 years (all but Q_{max}, which was not assessed after 1 year, and PVR, which was marginally significant at $P=.055$). Few severe complications were reported.[32] These results confirm that XPS is a good option for large prostates and suggest that continued advances in technique, such as vapoenucleation, can continue to increase the prostate volume for which PVP can be used.

Cost

One of the obstacles to adoption of PVP has been its perceived increase in cost relative to TURP. In the United Kingdom, in particular, a 2008 cost analysis suggested that PVP was not cost-effective when compared to TURP, and so the technology failed to achieve widespread use there.[33] Elsewhere, in countries where a single national health system does not make cost-effectiveness decisions, PVP has grown in usage and additional cost analyses have been performed. In general, PVP (and particularly XPS) has been found to be at least as cost effective as TURP; though the cost of surgery associated with laser usage is typically slightly higher than TURP, this is more than offset by the savings of a reduced hospital stay.[33–39] Due to the fact that a conventional TURP often requires an overnight stay in the hospital with CBI while PVP is an ambulatory procedure, the total cost in hospital resources to treat and recover a patient after PVP is often less than for a traditional TURP. Not all authors have found PVP to be more cost-effective than TURP. McCahy and colleagues performed a "bottom-up" rather than the typical "top–down" cost analysis and found that TURP was less expensive than PVP. This study, however, relied on the 120W system rather than the modern 180W and did not account for differences in prostate size that may have affected operative time.[40] Recent studies with XPS, by contrast, have found that both operative and postoperative costs are lower with PVP than with TURP, perhaps reflecting the increased speed of XPS and improved surgical technique as the technology matures.[38]

Conclusion

PVP has advanced significantly since the first 80W KTP systems were introduced. Subsequent improvements in power to 120 and then 180W lasers, along with increasing surgeon familiarity with this technology, have made PVP a powerful tool in the treatment of BPH. PVP is a safe and effective treatment and has been compared favorably to TURP, the current gold standard for BPH. Moreover, PVP is suitable for use in a broader range of patients, including those on anticoagulation for whom TURP would be contraindicated. As adoption continues to grow and new techniques are developed, including VIT and vapoenucleation, PVP will continue to be a mainstay of therapy for BPH.

References

1. Lee R, Gonzalez RR, Te AE. The evolution of photoselective vaporization prostatectomy (PVP): advancing the surgical treatment of benign prostatic hyperplasia. *World J Urol* 2006;**24**(4):405–9.
2. Zarrabi A, Gross AJ. The evolution of lasers in urology. *Ther Adv Urol* 2011;**3**(2):81–9.
3. Hecht J. A short history of laser development. *Appl Opt* 2010;**49**(25):F99.
4. Teichmann H-O, Herrmann TR, Bach T. Technical aspects of lasers in urology. *World J Urol* 2007;**25**(3):221–5.

5. Sountoulides P, Tsakiris P. The evolution of KTP laser vaporization of the prostate. *Yonsei Med J* 2008;**49**(2):189–99.
6. Hermanns T, Strebel DD, Hefermehl LJ, et al. Lithium triborate laser vaporization of the prostate using the 120 W, high performance system laser: high performance all the way? *J Urol* 2011;**185**(6):2241–7.
7. Brunken C, Munsch M, Tauber S, Schmidt R, Seitz C. The 532-nm 180-W (GreenLight®) laser vaporization of the prostate for the treatment of lower urinary tract symptoms: how durable is the new side-fire fiber with integrated cooling system? *Lasers Med Sci* 2014;**29**(3):1307–12.
8. Ko WJ, Choi BB, Kang HW, Rajabhandharaks D, Rutman M, Osterberg EC. Defining optimal laser-fiber sweeping angle for effective tissue vaporization using 180 W 532 nm lithium triborate laser. *J Endourol* 2012;**26**(4):313–7.
9. Zorn KC, Liberman D. GreenLight 180 W XPS photovaporization of the prostate: how I do it. *Can J Urol* 2011;**18**(5):5918–26.
10. Wosnitzer MS, Rutman MP. KTP/LBO laser vaporization of the prostate. *Urol Clin N Am* 2009;**36**(4):471–83.
11. Gomez Sancha F, Rivera VC, Georgiev G, Botsevski A, Kotsev J, Herrmann T. Common trend: move to enucleation—is there a case for GreenLight enucleation? Development and description of the technique. *World J Urol* 2015;**33**(4):539–47.
12. Netsch C, Bach T. Vaporization vs. enucleation techniques for BPO: do we have a standard? *Curr Opin Urol* 2015;**25**(1):45–52.
13. Elshal AM, Elkoushy MA, El-Nahas AR, et al. GreenLight™ laser (XPS) photoselective vapo-enucleation versus holmium laser enucleation of the prostate for the treatment of symptomatic benign prostatic hyperplasia: a randomized controlled study. *J Urol* 2015;**193**(3):927–34.
14. Azizi M, Tholomier C, Meskawi M, et al. Safety, perioperative, and early functional outcomes of vapor incision technique using the GreenLight XPS 180 W system: direct comparison with photoselective vaporization of the prostate. *J Endourol* 2017;**31**(1):43–9.
15. Misraï V, Faron M, Guillotreau J, et al. Assessment of the learning curves for photoselective vaporization of the prostate using GreenLight™ 180-Watt-XPS laser therapy: defining the intra-operative parameters within a prospective cohort. *World J Urol* 2014;**32**(2):539–44.
16. Malek RS, Kang HW, Peng YS, Stinson D, Beck MT, Koullick E. Photoselective vaporization prostatectomy: experience with a novel 180 W 532 nm lithium triborate laser and fiber delivery system in living dogs. *J Urol* 2011;**185**(2):712–8.
17. Bachmann A, Muir GH, Collins EJ, et al. 180-W XPS GreenLight laser therapy for benign prostate hyperplasia: early safety, efficacy, and perioperative outcome after 201 procedures. *Eur Urol* 2012;**61**(3):600–7.
18. Brunken C, Seitz C, Woo HH. A systematic review of experience of 180-W XPS GreenLight laser vaporisation of the prostate in 1640 men. *BJU Int* 2015;**116**(4):531–7.
19. Bachmann A, Tubaro A, Barber N, et al. 180-W XPS GreenLight laser vaporisation versus transurethral resection of the prostate for the treatment of benign prostatic obstruction: 6-month safety and efficacy results of a European multicentre randomised trial—the GOLIATH study. *Eur Urol* 2014;**65**(5):931–42.
20. Bachmann A, Tubaro A, Barber N, et al. A European multicenter randomized noninferiority trial comparing 180 W GreenLight XPS laser vaporization and transurethral resection of the prostate for the treatment of benign prostatic obstruction: 12-month results of the GOLIATH study. *J Urol* 2015;**193**(2):570–8.
21. Thomas JA, Tubaro A, Barber N, et al. A multicenter randomized noninferiority trial comparing GreenLight-XPS laser vaporization of the prostate and transurethral resection of the prostate for the treatment of benign prostatic obstruction: two-yr outcomes of the GOLIATH study. *Eur Urol* 2016;**69**(1):94–102.

22. Hahn RG. Fluid absorption in endoscopic surgery. *Br J Anaesth* 2005;**96**(1):8–20.
23. Wettstein MS, Poyet C, Grossmann NC, et al. Absorption of irrigation fluid during XPS™ GreenLight laser vaporization of the prostate: results from a prospective breath ethanol monitoring study. *World J Urol* 2016;**34**(9):1261–7.
24. Chung DE, Wysock JS, Lee RK, Melamed SR, Kaplan SA, Te AE. Outcomes and complications after 532 nm laser prostatectomy in anticoagulated patients with benign prostatic hyperplasia. *J Urol* 2011;**186**(3):977–81.
25. Woo HH, Hossack TA. Photoselective vaporization of the prostate with the 120-W lithium triborate laser in men taking coumadin. *Urology* 2011;**78**(1):142–5.
26. Shao IH, Hou C-P, Chen S-M, et al. The safety and efficacy of aspirin intake in photoselective vaporization laser treatment of benign prostate hyperplasia. *Clin Interv Aging* 2013;**8**:265–9.
27. Lee DJ, Rieken M, Halpern J, et al. Laser vaporization of the prostate with the 180-W XPS-Greenlight laser in patients with ongoing platelet aggregation inhibition and oral anticoagulation. *Urology* 2016;**91**:167–73.
28. Hueber P-A, Liberman D, Ben-Zvi T, et al. 180 W vs 120 W lithium triborate photoselective vaporization of the prostate for benign prostatic hyperplasia: a global, multicenter comparative analysis of perioperative treatment parameters. *Urology* 2013;**82**(5):1108–13.
29. Altay B, Erkurt B, Kiremit MC, Guzelburc V, Boz MY, Albayrak S. 180-W XPS GreenLight laser vaporization for benign prostate hyperplasia: 12-month safety and efficacy results for glands larger than 80 mL. *Lasers Med Sci* 2015;**30**(1):317–23.
30. Hueber P-A, Bienz MN, Valdivieso R, et al. Photoselective vaporization of the prostate for benign prostatic hyperplasia using the 180 Watt system: multicenter study of the impact of prostate size on safety and outcomes. *J Urol* 2015;**194**(2):462–9.
31. Meskawi M, Hueber P-A, Valdivieso R, et al. Multicenter international experience of 532 nm-laser photo-vaporization with Greenlight XPS in men with large prostates (prostate volume > 100 cc). *World J Urol* 2017. 1–7.
32. Stone BV, Chughtai B, Forde JC, Tam AW, Lewicki P, Te AE. Safety and efficacy of GreenLight XPS laser vapoenucleation in prostates measuring over 150 mL. *J Endourol* 2016;**30**(8):906–12.
33. Thomas JA, Tubaro A, Barber N, et al. The continuing story of the cost-effectiveness of photoselective vaporization of the prostate versus transuretheral resection of the prostate for the treatment of symptomatic benign prostatic obstruction. *Value Health* 2015;**18**(4):376–86.
34. Ruszat R, Sulster T, Seifert HH, et al. Photoselective vaporisation (PVP) vs. transurethral electroresection of the prostate (TURP): a comparing cost analysis. *Eur Urol Suppl* 2006;**5**(2):271.
35. Stovsky MD, Griffiths RI, Duff SB. A clinical outcomes and cost analysis comparing photoselective vaporization of the prostate to alternative minimally invasive therapies and transurethral prostate resection for the treatment of benign prostatic hyperplasia. *J Urol* 2006;**176**(4):1500–6.
36. Liatsikos E, Kyriazis I, Kallidonis P, Sakellaropoulos G, Maniadakis N. Photoselective GreenLight™ laser vaporization versus transurethral resection of the prostate in Greece: a comparative cost analysis. *J Endourol* 2012;**26**(2):168–73.
37. Benejam-Gual JM, Sanz-Granda A, Budía A, Extramiana J, Capitán C. Multicenter study on costs associated with two surgical procedures: GreenLight XPS 180 W versus the gold standard transurethral resection of the prostate. *Actas Urol Esp* 2014;**38**(6):373–7.
38. Shah A, Welliver C, Chang-Kit L. PD24-02 A Cost analysis of 180W photoselective vaporization of the prostate (PVP) vs transurethral resection of the prostate (TURP). *J Urol* 2016;**195**(4):e511.

39. Goh AC, Gonzalez RR. Photoselective laser vaporization prostatectomy versus transurethral prostate resection: a cost analysis. *J Urol* 2010;**183**(4):1469–73.
40. McCahy P, Cheng K, Paul E, Gleeson J. The cost of photoselective vaporization of the prostate compared to transurethral resection of the prostate: experience in a large public Australian teaching centre. *J Clin Urol* 2014;**7**(5):323–37.

Further Reading

1. Ben-Zvi T, Hueber P-A, Liberman D, Valdivieso R, Zorn KC. GreenLight XPS 180 W vs HPS 120 W laser therapy for benign prostate hyperplasia: a prospective comparative analysis after 200 cases in a single-center study. *Urology* 2013;**81**(4):853–8.
2. Rieken M, Bonkat G, Müller G, et al. The effect of increased maximum power output on perioperative and early postoperative outcome in photoselective vaporization of the prostate. *Lasers Surg Med* 2013;**45**(1):28–33.

Holmium Laser Ablation of the Prostate

10

H. Gregory Moore, Dominique Thomas, Bilal Chughtai
Weill Cornell Medicine, New York, NY, United States

Introduction

Since the first urologic uses in the 1980s, laser has become an important component of surgical management of benign prostatic hyperplasia (BPH) with mounting evidence of their advantages from treating patients on anticoagulants to improved postoperative recovery times. Laser use has evolved rapidly to include novel technologies and existing laser technologies, each laser able to adjust to different settings and different solid-state alloys enabling different wavelengths and energies to penetrate to different tissue depths. The suite of solid-state lasers in use today is neodymium: yttrium aluminum garnet (Nd:YAG), holmium (Ho:YAG), and thulium (Tm:YAG), and photovaporization of the prostate (PVP) also known as GreenLight or potassium titanyl phosphate (KTP). This alphabet soup of laser modalities affords the surgeon a wide variety of choices, allowing for individualized selection of the optimum procedure for the patient, as each laser exacts a unique effect on tissue.

This chapter aims to impart a practical knowledge of the basic features and applications of the Ho:YAG energy source in treating BPH. There are three categories of prostate surgery: ablation (superheating intracellular water of prostatic tissue to above boiling point causing it to vaporize along with the surrounding tissue), resection (excision of fragments of tissue and irrigation out of the bladder), and enucleation (dissection of the hyperplastic tissue en bloc and morcellation of dissected tissue). Here, the focus will be on the holmium laser in vaporizing or ablating the prostate as well as a review of literature findings on outcomes relative to TURP as well as to other novel methods.

Laser use in surgery began in 1962 in the realm of cardiothoracic surgery, 2 years after the invention of the first laser by Dr. Theodore H. Maiman.[1] Urologic use of lasers began in the 1980s in fragmenting kidney stones.[2] In 1990, Dr. Anthony Costello used a neodymium: yttrium aluminum garnet (Nd:YAG) laser in a patient for the first time to treat BPH, visual ablation of the prostate (VLAP).[3] The Nd:YAG laser was promptly supplanted by the holmium:YAG (Ho:YAG) laser in 1995 when Drs. Mark Fraundorfer and Peter Gilling experimented with combination treatment of BPH where they used a Ho:YAG laser to bore holes into the prostate and used a Nd:YAG laser to coagulate.[4] They found that Ho:YAG laser had superior properties to prostate ablation and could be used independently of the Nd:YAG laser thus establishing holmium laser ablation of the prostate (HoLAP). With two proven approaches, the benefits of laser ablation were recognized and jump-started the field of laser treatment for BPH.

A Comprehensive Guide to the Prostate. https://doi.org/10.1016/B978-0-12-811464-3.00010-6

Physics and History

Laser deliver a focused beam of light at a single wavelength to target tissues which convert light energy into heat energy through absorption. The light energy is absorbed by molecules called chromophores resident in the tissue resulting in an increase in temperature. The depth to which a laser is able to penetrate is dependent on wavelength and the type of tissue the laser is entering. In human prostate tissue, hemoglobin and water act as chromophores, each absorbing light optimally at slightly different wavelengths. Another consideration for laser technologies is reflection. When the beam encounters nonhomogenous tissue, the boundary between one tissue medium and another scatters the incident beam, causing potential effects on surrounding tissue. Light of shorter wavelengths scatter more than those of longer wavelengths. Lasers can either be front-firing, where the radiation is emitted from the tip, or side firing, where the laser radiation is direct tangentially to the tip of the fiber (typically 70 degrees).

The Nd:YAG laser has an intermediate wavelength of 1064 nm and is absorbed by both hemoglobin and water. This allows the laser to easily pass through the liquid irrigant and penetrate deeper into tissue to about 5 mm. However, this imparts thermal injury to tissues that will undergo delayed necrosis and sloughing postoperatively. Accordingly, patients experienced more prolonged catheterization and dysuria. Several years later, novel technologies able to deliver higher power output were successfully developed that had more favorable effects on prostate tissue.

The Ho:YAG laser has a longer wavelength (2120 nm) and is absorbed exclusively by water, penetrating to a tissue depth of only 0.4 mm. The selective absorption by water creates a vapor bubble at the tip of the fiber able to deliver a high amount of energy to the adjacent tissue without diffusing any energy through water and quickly raises the temperature above boiling, instantly vaporizing the intracellular fluid and removing the tissue.[5]

Surgical Indications

Ideal surgical candidates for HoLAP include patients with prostates less than 60 cc with a 60 W Ho:YAG laser. Newer research provides evidence that prostates with volumes of greater than 80 ccs can be safely ablated with more powerful 100 W Ho:YAG laser, yet other research reinforces the recommendation for exclusive use of HoLAP in smaller prostate adenomas.[6–8] Successful ablation has been performed on prostates larger than 200 cc. Because HoLAP requires longer operation times for larger prostates, longer than 200 min for exceptionally large prostates, special consideration should be taken when weighing treatment approaches to men with large prostates.[9] There is a steep learning curve with large prostates.

One of the greatest benefits of the holmium laser approach to BPH treatment is the indication for use in patients on anticoagulant therapy as well as avoiding TUR syndrome, caused by the absorption of irrigant during surgery causing hyponatremia, vomiting, dizziness, and seizure as well as reducing the requirement for blood transfusions[10,11].

Surgical Technique

Preparation

The goal of the procedure is to systematically remove all prostatic tissue until the prostatic capsule is reached. The holmium laser is designed to quickly vaporize tissue without carbonizing or charring it when held in close approximation to the tissue. Once the tissue is vaporized, irrigant will fill in the area that was occupied by the now-vaporized tissue and will dissipate the energy of the laser, preserving the deeper tissue until the laser is brought closer to the new tissue to be vaporized meaning the laser does not need to be kept in constant motion to avoid overtreating a single area. Notably, the laser should not be inserted into the tissue directly as the tissue that adheres to jacket of the fiber will be heated by the laser and can transmit that heat to the jacket itself, causing material breakdown.

The tissue is then cut posteriorly using a side fire, with the fiber being close to the prostate.

While the fiber is in a fixed position, the end of the scope with the camera is lifted up. The scope is then pulled from the bladder neck to the veru. The urologist will repeat this action until the capsule fibers have been identified.

The scope and fiber can be advanced to the bladder neck once the capsule is visible. From the incision at the leave of the capsule, the aiming bean can be directed laterally allowing the urologist to move the tissue to the side. The aiming bean stays in this fixed position and is then moved laterally to maximize absorption and heat generation. This results in a more efficient ablative technique.

The procedure begins proximally at the vesicle neck and the firing tip of the laser is gradually withdrawn more distally until reaching the verumontanum. This is so because ablated, coarse tissue margins as well as collapsed portions of the prostate that fall into the operating field will obstruct the view of the resectoscope as well as impede flow of irrigant. As detailed above, the surgeon will ablate prostate tissue in slow sweeping motions. If portions of the prostate do collapse into the operating field, it is important to ablate them, maintaining the proximal-distal motion intact.

Postoperative Management

After surgery, a two-way 20F Foley indwelling catheter is inserted for 1–2 days, though oftentimes patients can be discharged within 24 h without a catheter and mild dysuria treatable with analgesics as ablated tissue sloughs off.

Duration of symptoms after holmium ablation surgery such as retrograde ejaculation, urethral strictures, urinary retention, urinary tract infection, and stress and urge incontinence is comparable to TURP. Transient symptoms can be managed with analgesics, antimuscarinic drugs, anticholinergic medications, and antimicrobials depending on the situation. Occasionally procedural intervention may be required.[12] Importantly, the degree of bladder function prior to surgery will determine the degree and duration of irritative bladder symptoms after surgery.

Outcomes

Numerous studies have been conducted to evaluate the outcomes of HoLAP procedures since its introduction in the 1990s, beginning with Drs. Fraundorfer and Gilling in 1995 and published a 7-year follow-up that reported durable results from the initial cohort of patients receiving HoLAP treatment, with a retreatment rate of 15%. Holimum laser technology remains an attractive option because it reduces intraoperative bleeding and catheterization time and virtually eliminates dysuria immediately after surgery.

HoLAP compares favorably to PVP as well. A 4-year follow-up study by Elmansi et al. concluded that PVP and HoLAP had durable improvement in IPSS, PVR, and sexual function. However, Elshal et al. found that both HoLAP and PVP did not achieve significant improvement 1 year after surgery. HoLEP was found to improve International Index of Erectile Function scores postoperatively despite finding that HoLEP-treated patients had significantly increased onset of retrograde ejaculation compared to HoLAP-treated patients.[13]

HoLAP has been compared to TURP in only one randomized controlled trial conducted in 1999 on 36 patients with a follow-up of 1 year. The authors found that the HoLAP procedures took roughly 20 min longer than TURP (75 vs. 56 min), resulted in shorter catheterization times (1.7 vs. 1.2 days), and smaller reductions in AUA score (20–7 vs. 24.1 to 5). There was no significant difference detected between the groups regarding potency or ejaculatory status and Q max.[14] In 2013, long-term outcomes for PVP and HoLAP were reported to be similar, both with reoperations rates for bladder obstruction around 20%. Most recently, Jayadevan et al. compared 100W VeraPulse HoLAP to PVP, TURP, and PlasmaButton vaporization, finding HoLAP comparable to TURP.[15]

Learning Curve

Compared to other surgical techniques for BPH, HoLAP can be taught to residents and attending physicians with ease, as the learning curve is flatter than for other approaches such as HoLEP or PVP. Holmium lasers are used in other urologic procedures such as stone fragmentation and urethral strictures, giving trainees exposure to the modality in other contexts and enabling easier adoption. Herrick et al., compared operative times for residents in their third year of training to attending surgeons and found were only 11 min slower (57 vs. 46 min) than experienced HoLAP operators and both achieved similar outcomes (postvoid residual and IPSS) and reoperation rates. This demonstrating HoLAP can be applied safely and effectively even in when utilized by novice practitioners.[16]

Summary

Holmium laser technology ushered in a new era of prostate surgery with HoLAP, HoLRP, and HoLEP all affording outcomes that rival that of the gold standard (TURP). The Ho:YAG modality penetrates to a shallow tissue depth to vaporize tissue and is able to coagulate even deeper tissues through head conduction minimizing blood loss as well as shortened catheterization time and reducing postoperative dysuria. Although HoLAP requires longer operative times, the technique is highly adoptable and user-friendly compared to enucleation.

References

1. Choy D. History of lasers in medicine. *Thorac Cardiovasc Surg* 1988;**36**(S 2):114–7.
2. Zarrabi A, Gross AJ. The evolution of lasers in urology. *Ther Adv Urol* 2011;**3**(2):81–9.
3. Costello AJ, Bowsher WG, Bolton DM, Braslis KG, Burt J. Laser ablation of the prostate in patients with benign prostatic hypertrophy. *Br J Urol* June 1992;**69**(6):603–8.
4. Gilling PJ, Cass CB, Malcolm AR, Fraundorfer MR. Combination holmium and Nd:YAG laser ablation of the prostate: initial clinical experience. *J Endourol* April 1995;**9**(2): 151–3.
5. Teichmann HO, Herrmann TR, Bach T. Technical aspects of lasers in urology. *World J Urol* June 2007;**25**(3):221–5.
6. Gilling PJ, Cass CB, Cresswell MD, Malcolm AR, Fraundorfer MR. The use of the holmium laser in the treatment of benign prostatic hyperplasia. *J Endourol* October 1996;**10**(5):459–61.
7. Elzayat EA, Al-Mandil MS, Khalaf I, Elhilali MM. Holmium laser ablation of the prostate versus photoselective vaporization of prostate 60 cc or less: short-term results of a prospective randomized trial. *J Urol* July 2009;**182**(1):133–8.
8. Barski D, Richter M, Winter C, et al. Holmium laser ablation of the prostate (HoLAP): intermediate-term results of 144 patients. *World J Urol* October 2013;**31**(5):1253–9.
9. Kumar SM. Rapid communication: holmium laser ablation of large prostate glands: an endourologic alternative to open prostatectomy. *J Endourol* June 2007;**21**(6):659–62.
10. Kollmeier MA, Stock RG, Cesaretti J, Stone NN. Urinary morbidity and incontinence following transurethral resection of the prostate after brachytherapy. *J Urol* March 2005;**173**(3):808–12.
11. Patel H, Peddada AV, Zimmern PE, Hernandez R, Kagan R. Risk of incontinence with transurethral resection of the prostate after radiation therapy for prostate cancer. *J Surg Oncol* 1997;**64**(2):127–9.
12. Elshal AM, Elmansy HM, Elhilali MM. Two laser ablation techniques for a prostate less than 60 mL: lessons learned 70 months after a randomized controlled trial. *Urology* 2013;**82**(2):416–24.
13. Elshal AM, Elmansy HM, Elkoushy MA, Elhilali MM. Male sexual function outcome after three laser prostate surgical techniques: a single center perspective. *Urology* November 2012;**80**(5):1098–104.
14. Mottet N, Anidjar M, Bourdon O, et al. Randomized comparison of transurethral electroresection and holmium: YAG laser vaporization for symptomatic benign prostatic hyperplasia. *J Endourol* March 1999;**13**(2):127–30.
15. Jayadevan R, Stensland K, Baldwin G, Ali A, Hall S, Palese M. Comparison of transurethral resection of prostate, plasma button (Tm) transurethral vaporization of prostate, Greenlight (Tm) 180-W XPS photoselective vaporization of prostate, and versapulse (R) 100-W holmium laser ablation of prostate. *J Urol* April 2014;**191**(4):E790.
16. Herrick BW, Yap RL. It is safe to teach residents laser prostate surgery in the private practice setting. *Urology* March 2013;**81**(3):629–32.

Holmium Enucleation of the Prostate

11

H. Gregory Moore, Dominique Thomas, Bilal Chughtai
Weill Cornell Medicine, New York, NY, United States

Introduction

There are several surgical approaches to treating benign prostatic hyperplasia (BPH), divisible into three groups: ablation, resection, and enucleation. Ablation entails vaporizing tissue by heating the tissue rapidly to boiling point. Resection removes small portions of the prostate. Finally, enucleation involves dissecting around the anatomical plane of the capsule of the prostate adenoma and then removing it en bloc. Enucleation is an evolved hybrid, inheriting its conceptual approach, dissection along anatomically demarcated incision planes, from open prostatectomy (OP) while applying a transurethral holmium laser:yttrium aluminum garnet (Ho:YAG) laser. Enucleation has been heralded as the closest approximation, if not equal, to the gold standard of BPH treatment: transurethral resection of the prostate (TURP). This chapter aims to examine the history, the technique, and outcomes of holmium laser enucleation of the prostate (HoLEP).

Laser Properties

Lasers exert their effect on tissues by transmitting light energy to heat energy through interaction with light-absorbing molecules called chromophores. The two chromophores relevant to the human prostate are hemoglobin and water. The Ho:YAG laser operates at a wavelength of 2120 nm making tissue water the primary chromophore. Intracellular fluid is rapidly brought to boiling point, vaporizes, and destroys the tissue.[1] Additionally, there is a complementary vapor bubble effect at the tip of the glass fiber. This bubble forms when laser radiation exits the fiber and vaporizes the water around the tip of the fiber. When the fiber tip and bubble come into contact with tissue, the energy penetrates to a depth of 0.4 mm and vaporizes the intervening tissue. Heat conduction through heterogeneous materials will coagulate deeper tissues without vaporizing them, helping with hemostasis.[2] The holmium laser emits a pulsatile beam, which enables better visualization of the structures being dissected compared to other continuous lasers such as neodymium (Nd:YAG) and thulium (Tm:YAG). The holmium energy source can be utilized in three BPH treatment approaches, ablation (holmium laser ablation of the prostate [HoLAP]), resection (holmium laser resection of the prostate [HoLRP]), and enucleation (HoLEP).

A Comprehensive Guide to the Prostate. https://doi.org/10.1016/B978-0-12-811464-3.00011-8

Evolution of Holmium Laser Enucleation of the Prostate

Though Ho:YAG laser use in urology dates back to the late 1980s for use in fragmenting renal calculi, the first application of the holmium energy source to treat BPH was reported by Drs. Gilling and Chun in 1995.[3,4] The authors first reported strong clinical outcomes using a dual approach of Nd:YAG for incision and Ho:YAG for ablation but subsequently abandoned the use of Nd:YAG because holmium ablation offered superior features more suitable to application in treating BPH. HoLAP was compared in a randomized controlled trial (RCT) to TURP and found to be superior in some ways (shorter catheterization time), comparable with similar improvements in Qmax and International Prostate Symptom Score (IPSS) scores and no difference in the postoperative sexual dysfunction and inferior in other parameters, namely the procedure took 20 min longer on average than TURP.[5] This limits this technique to prostates that are less than 60 ccs. Holmium ablation has decreased in popularity but the concept remains valid and has been adopted by newer technologies including photovaporization of the prostate (PVP), thulium vaporization of the prostate, and higher power holmium lasers (100 and 120 W).[6,7]

The next iteration of holmium laser application in BPH was HoLRP, developed to reduce operative times to durations that more closely resemble TURP. HoLRP relies on using a Ho:YAG laser to incise pieces of prostatic tissue piece by piece until exposing the surgical capsule rather than ablating half-millimeter layers of tissue. The success of HoLRP has been more thoroughly studied and a metaanalysis by Tooher et al. HoLRP is as effective in improving Qmax and quality of life scores as TURP.[8] However, HoLRP patients experienced shorter catheterization duration and decreased transfusion rates when compared to the TURP group.[9] Additionally, Frauendorfer et al. found that HoLRP was nearly one-fourth less expensive than TURP and in less than 100 cases the initial capital investment in a Ho:YAG laser could be recouped.[10] Like HoLAP, the primary weakness of HoLRP was operative time, which was found to be as high as 30% longer than TURP. Fragmenting the lobes of the prostate with repeated holmium incision, while virtually bloodless, takes significantly longer than the electrocautery applied in TURP. Accordingly, HoLRP has fallen out of favor for treating BPH except in certain scenarios such as reoperation for adenomatous regrowth.

HoLEP allows the surgeon to completely dissect out the adenoma by using the Ho:YAG laser to incise along the capsule to shell out the lobes of the adenoma en bloc. The shelled-out lobes are then deposited into the bladder for morcellation. This technique relies on an entirely new component of technology, the transurethral power morcellator, to avoid having to remove bits of the prostate piecemeal through the urethra. A morcellator uses reciprocating blades to simultaneously shred and suction. First reported by Fraundorfer and Gilling in 1998 on a series of 14 patients where both suprapubic and transurethral morcellation was used. The results demonstrated favorable outcomes for patients with average glad size of 98.6 ccs and with an average surgical duration of 98 min. Twelve of the patients were discharged the next day without a catheter and at 1 month, Qmax was 25.2 mL/s and mean IPSS was 7.2.[11] In the same year, a report on a larger group of 64 patients was published that reported similarly, compelling outcomes: greater than 90% discharged next day catheter-free and Qmax of 23.4 mL/s, and a reduction in IPSS from 23.0 to 8.6 after 1 month.[12,13]

The size of the prostate, while potentially extending operation time, does not affect perioperative clinical outcomes such as length of stay, Qmax, postvoid residual, and catheterization time.[14–16]

Technique

Bladder Neck Incisions

The resectoscope is inserted and the important landmarks are identified. The purpose of this step is to assess the pathology and the surface area of the uretic orifices where the bladder neck incisions are made. These incisions can occur at the 5 o'clock and 7 o'clock positions.

Proximal to the verumontanum, the ends of the incisions are made to the bladder neck with a transverse incision. Toward the bladder neck, in a retrograde fashion dissection of the median lobe is on the capsule. The resectoscope is used to manipulate the median lobe all the way up into the bladder neck. The adenoma is then dissected off the capsule.

From the initial bladder neck, the later lobes are incised at the 2 and 10 o'clock positions.

A nephroscope is used for morcellation. This offsets the telescope lens.

This procedure has been described in detail by Lingeman et al.[17]

Randomized Trials

In the nearly 20 years of HoLEP experience, a wealth of randomized trials have been conducted comparing HoLEP to TURP and OP procedures as well as newer techniques such as PVP and Tm:YAG enucleation (thulium laser vapoenucleation [ThuVEP]).[14,18–34]

A systematic metaanalysis of eight RCTs comparing HoLEP to TURP in 2014 found that there was no significant difference in IPSS score at 3 and 6 months between techniques. At 12 months, the authors found improved IPSS scores in HoLEP compared to TURP and higher Qmax values for HoLEP at 3 and 12 months compared to TURP, concluding that HoLEP was superior to TURP when considering clinical outcomes but did not evaluate postoperative complications. Another metaanalysis comparing six RCTSs found that HoLEP procedures resulted in significantly less blood loss, catheterization time, and blood transfusions. However, operative times and postoperative dysuria were significantly higher in HoLEP compared to TURP.[18]

With HoLEP's extensive track record of proven equivalence and superior safety, efficacy, and durability relative to TURP, HoLEP is considered by many to be the novel gold standard.[35–38] To bolster that assertion, HoLEP has begun to be compared to newer entrants to the transurethral laser surgery market such as GreenLight/PVP/KTP and PlasmaButton in the age of laser urologic surgery. An RCT comparing HoLEP to GreenLight laser photoselective vapoenucleation (PVP) found that improvement in PVR was comparable between the procedures. However, the percentage reduction in prostate size (74.3% vs. 43.1%) and Qmax at 12 months were significantly higher in the HoLAP group.[30] Several retrospective studies had similar findings when compared to

PVP.[39,40] Finally, a comparison with ThuVEP found no difference in Q max, urinary retention, operative time, or complications after 6 months.[41]

Learning Curve

HoLEP is a complex procedure that is applied to a wider range of prostate sizes leading to an infamous reputation of having a steep learning curve. There is a growing body of research on how to impart the technical expertise from one surgeon to another because of this widely acknowledged barrier. Brunckhorst et al. studied a single surgeon over 7 years of HoLEP surgery. They found that the parameters stabilized around 40–60 cases.[42] Fortunately, during the learning curve, outcomes were not affected.[43] In the United Kingdom, a Holmium User Group (HUG) was established in 2006 to tackle the perceived difficulty and prolonged training time required to learn HoLEP. Mentors were required to have completed more than 100 HoLEP procedures and scheduled visits with mentee surgeons to give surgeons the chance to experience the requisite number of surgeries to achieve proficiency. With only 6 mentors, they were able to train over 30 surgeons in a few short years.[44] In the early stages, simulation has also been utilized to train participants during a course hosted by HUG, an innovative solution to the high learning barrier.[45] Ultimately, HoLEP is a highly reproducible procedure, with the proper resources, and experienced mentorship, residents, and attending surgeons alike can become expert HoLEP practitioners.[46,47]

Cost

HoLEP is 25% less expensive than TURP with reduced nursing staff demand due to decreased catheterization time, bladder irrigation, and clot retention. Availability of reusable laser fibers and morcellator blades affords HoLEP even more cost savings compared to other surgical approaches. In a prospective study following 63 consecutive patients during either HoLEP or OP surgery and throughout their postoperative hospital stays found that the mean perioperative cost of HoLEP was significantly less expensive ($2919.4 vs. $3556.30 dollars). However, costs were higher for HoLEP intraoperatively because of longer surgical duration and higher material costs. Overall, the cost of HoLEP was 9.6% less than OP largely because of significantly reduced length of stay after HoLEP. Another study similar study was conducted more recently in a developing country compared HoLEP to transvesical OP over the course of 3 months. HoLEP operated patients experienced lower transfusion rates (2.1% vs. 26.1%) and shorter mean hospital stays (2 vs. 9 days) and comparable Qmax, postvoid residual volume, and % PSA reduction. HoLEP was found to be less expensive but not significantly.[48] As suggested by the authors, increased experience with HoLEP decreased costs over time. Moreover, when compared to photoselective vapoenucleation (GreenLight) in a randomized controlled setting, HoLEP had lower cost per procedure.[30] When taken together, these studies demonstrate the favorable economics of HoLEP globally especially when it comes to large prostates (larger than 80 g).

Summary

There are numerous surgical approaches for addressing BPH. Although individual preference, training, and circumstance tend to determine the technique surgeons use, HoLEP has an ever growing body of high-quality evidence in achieving durable outcomes. RCTs comparing HoLEP to newer laser-based approaches as well as pitting it against OP and TURP support its place as a cost-effective, teachable approach to BPH.

References

1. Teichmann HO, Herrmann TR, Bach T. Technical aspects of lasers in urology. *World J Urol* June 2007;**25**(3):221–5.
2. Jacques SL. Laser tissue interactions – photochemical, photothermal, and photomechanical. *Surg Clin N Am* June 1992;**72**(3):531–58.
3. Chun SS, Razvi HA, Denstedt JD. Laser prostatectomy with the holmium: YAG laser. *Tech Urol* Winter 1995;**1**(4):217–21.
4. Gilling PJ, Cass CB, Malcolm AR, Fraundorfer MR. Combination holmium and Nd:YAG laser ablation of the prostate: initial clinical experience. *J Endourol* April 1995;**9**(2):151–3.
5. Mottet N, Anidjar M, Bourdon O, et al. Randomized comparison of transurethral electroresection and holmium: YAG laser vaporization for symptomatic benign prostatic hyperplasia. *J Endourol* March 1999;**13**(2):127–30.
6. Zhang X, Shen P, He Q, et al. Different lasers in the treatment of benign prostatic hyperplasia: a network meta-analysis. *Sci Rep* 2016;**6**.
7. Tholomier C, Valdivieso R, Hueber P-A, Zorn KC. Photoselective laser ablation of the prostate: a review of the current 2015 tissue ablation options. *Can J Urol* 2015;**22**(5 Suppl. 1):45–52.
8. Tooher R, Sutherland P, Costello A, Gilling P, Rees G, Maddern G. A systematic review of holmium laser prostatectomy for benign prostatic hyperplasia. *J Urol* May 2004;**171**(5):1773–81.
9. Das A, Kennett K, Fraundorfer M, Gilling P. Holmium laser resection of the prostate (HoLRP): 2-year follow-up data. *Tech Urol* December 2001;**7**(4):252–5.
10. Fraundorfer MR, Gilling PJ, Kennett KM, Dunton NG. Holmium laser resection of the prostate is more cost effective than transurethral resection of the prostate: results of a randomized prospective study. *Urology* 2001;**57**(3):454–8.
11. Fraundorfer MR, Gilling PJ. Holmium: YAG laser enucleation of the prostate combined with mechanical morcellation: preliminary results. *Eur Urol* 1998;**33**(1):69–72.
12. Gilling PJ, Kennett K, Das AK, Thompson D, Fraundorfer MR. Holmium laser enucleation of the prostate (HoLEP) combined with transurethral tissue morcellation: an update on the early clinical experience. *J Endourol* October 1998;**12**(5):457–9.
13. Tyson MD, Lerner LB. Safety of holmium laser enucleation of the prostate in anticoagulated patients. *J Endourol* 2009;**23**(8):1343–6.
14. Kuntz RM, Lehrich K, Ahyai S. Does perioperative outcome of transurethral holmium laser enucleation of the prostate depend on prostate size? *J Endourol* March 2004;**18**(2):183–8.
15. Hettiarachchi JA, Samadi AA, Konno S, Das AK. Holmium laser enucleation for large (greater than 100 mL) prostate glands. *Int J Urol* May 2002;**9**(5):233–6.
16. Lee MH, Yang HJ, Kim DS, Lee CH, Jeon YS. Holmium laser enucleation of the prostate is effective in the treatment of symptomatic benign prostatic hyperplasia of any size including a small prostate. *Korean J Urol* November 2014;**55**(11):737–41.

17. Lingeman RLK, Ryan FP, Samuel CK, Tibério Jr MS, Mostafa ME, James E. Holmium laser enucleation of the prostate (HoLEP): a technical update. *World J Surg Oncol* June 06, 2003;**1**(1):6.
18. Tan AH, Gilling PJ, Kennett KM, Frampton C, Westenberg AM, Fraundorfer MR. A randomized trial comparing holmium laser enucleation of the prostate with transurethral resection of the prostate for the treatment of bladder outlet obstruction secondary to benign prostatic hyperplasia in large glands (40 to 200 grams). *J Urol* October 2003;**170**(4 Pt 1):1270–4.
19. Kuntz RM, Lehrich K, Ahyai SA. Holmium laser enucleation of the prostate versus open prostatectomy for prostates greater than 100 grams: 5-year follow-up results of a randomised clinical trial. *Eur Urol* January 2008;**53**(1):160–8.
20. Naspro R, Bachmann A, Gilling P, et al. A review of the recent evidence (2006-2008) for 532-nm photoselective laser vaporisation and holmium laser enucleation of the prostate. *Eur Urol* June 2009;**55**(6):1345–57.
21. Montorsi F, Naspro R, Salonia A, et al. Holmium laser enucleation versus transurethral resection of the prostate: results from a 2-center, prospective, randomized trial in patients with obstructive benign prostatic hyperplasia. *J Urol* November 2004;**172**(5 Pt 1): 1926–9.
22. Briganti A, Naspro R, Gallina A, et al. Impact on sexual function of holmium laser enucleation versus transurethral resection of the prostate: results of a prospective, 2-center, randomized trial. *J Urol* May 2006;**175**(5):1817–21.
23. Gupta N, Sivaramakrishna KR, Dogra PN, Seth A. Comparison of standard transurethral resection, transurethral vapour resection and holmium laser enucleation of the prostate for managing benign prostatic hyperplasia of >40 g. *BJU Int* January 2006;**97**(1):85–9.
24. Naspro R, Suardi N, Salonia A, et al. Holmium laser enucleation of the prostate versus open prostatectomy for prostates >70 g: 24-month follow-up. *Eur Urol* September 2006;**50**(3):563–8.
25. Neill MG, Gilling PJ, Kennett KM, et al. Randomized trial comparing holmium laser enucleation of prostate with plasmakinetic enucleation of prostate for treatment of benign prostatic hyperplasia. *Urology* November 2006;**68**(5):1020–4.
26. Wilson LC, Gilling PJ, Williams A, et al. A randomised trial comparing holmium laser enucleation versus transurethral resection in the treatment of prostates larger than 40 grams: results at 2 years. *Eur Urol* September 2006;**50**(3):569–73.
27. Ahyai SA, Lehrich K, Kuntz RM. Holmium laser enucleation versus transurethral resection of the prostate: 3-year follow-up results of a randomized clinical trial. *Eur Urol* 2007;**52**(5):1456–64.
28. Gilling PJ, Wilson LC, King CJ, Westenberg AM, Frampton CM, Fraundorfer MR. Long-term results of a randomized trial comparing holmium laser enucleation of the prostate and transurethral resection of the prostate: results at 7 years. *BJU Int* February 2012;**109**(3):408–11.
29. Chen YB, Chen Q, Wang Z, et al. A prospective, randomized clinical trial comparing plasmakinetic resection of the prostate with holmium laser enucleation of the prostate based on a 2-year followup. *J Urol* January 2013;**189**(1):217–22.
30. Elshal AM, Elkoushy MA, El-Nahas AR, et al. GreenLight laser (XPS) photoselective vapo-enucleation versus holmium laser enucleation of the prostate for the treatment of symptomatic benign prostatic hyperplasia: a randomized controlled study. *J Urol* March 2015;**193**(3):927–34.
31. Fayad AS, Elsheikh MG, Zakaria T, et al. Holmium laser enucleation of the prostate versus bipolar resection of the prostate: a prospective randomized study. "Pros and cons". *Urology* November 2015;**86**(5):1037–41.

32. Sun N, Fu Y, Tian T, et al. Holmium laser enucleation of the prostate versus transurethral resection of the prostate: a randomized clinical trial. *Int Urol Nephrol* July 2014;**46**(7):1277–82.
33. Basic D, Stankovic J, Potic M, Ignjatovic I, Stojkovic I. Holmium laser enucleation versus transurethral resection of the prostate: a comparison of clinical results. *Acta Chir Iugosl* 2013;**60**(1):15–20.
34. Zhang F, Shao Q, Herrmann TR, Tian Y, Zhang Y. Thulium laser versus holmium laser transurethral enucleation of the prostate: 18-month follow-up data of a single center. *Urology* April 2012;**79**(4):869–74.
35. van Rij S, Gilling PJ. In 2013, holmium laser enucleation of the prostate (HoLEP) may be the new 'gold standard'. *Curr Urol Rep* December 2012;**13**(6):427–32.
36. Vincent MW, Gilling PJ. HoLEP has come of age. *World J Urol* 2015;**33**(4):487–93.
37. Elzayat EA, Habib EI, Elhilali MM. Holmium laser enucleation of the prostate: a size-independent new "gold standard". *Urology* November 2005;**66**(5 Suppl.):108–13.
38. Gnessin E, Mandeville JA, Lingeman JE. An update on holmium laser enucleation of the prostate and why it has stood the test of time. *Curr Opin Urol* January 2011;**21**(1):31–5.
39. Jaeger CD, Mitchell CR, Mynderse LA, Krambeck AE. Holmium laser enucleation (HoLEP) and photoselective vaporisation of the prostate (PVP) for patients with benign prostatic hyperplasia (BPH) and chronic urinary retention. *BJU Int* February 2015;**115**(2):295–9.
40. Elshal AM, Elkoushy MA, Elmansy HM, Sampalis J, Elhilali MM. Holmium:YAG transurethral incision versus laser photoselective vaporization for benign prostatic hyperplasia in a small prostate. *J Urol* January 2014;**191**(1):148–54.
41. Netsch C, Becker B, Tiburtius C, et al. Thulium vapoenucleation of the prostate versus holmium laser enucleation of the prostate for the treatment of benign prostatic obstruction: 6-month safety and efficacy results of a prospective randomized trial. *Eur Urol Suppl* 2017;**16**(3):e512–3.
42. Brunckhorst O, Ahmed K, Nehikhare O, Marra G, Challacombe B, Popert R. Evaluation of the learning curve for holmium laser enucleation of the prostate using multiple outcome measures. *Urology* October 2015;**86**(4):824–9.
43. Elzayat EA, Elhilali MM. Holmium laser enucleation of the prostate (HoLEP): long-term results, reoperation rate, and possible impact of the learning curve. *Eur Urol* November 2007;**52**(5):1465–71.
44. Aho T, Herrmann TR. Description of a modular mentorship programme for holmium laser enucleation of the prostate. *World J Urol* April 2015;**33**(4):497–502.
45. Aydin A, Ahmed K, Brewin J, Khan MS, Dasgupta P, Aho T. Face and content validation of the prostatic hyperplasia model and holmium laser surgery simulator. *J Surg Educ* May–June 2014;**71**(3):339–44.
46. Kim SC, Matlaga BR, Kuo RL, et al. Holmium laser enucleation of the prostate: a comparison of efficiency measures at two institutions. *J Endourol* June 2005;**19**(5):555–8.
47. El-Hakim A, Elhilali MM. Holmium laser enucleation of the prostate can be taught: the first learning experience. *BJU Int* December 2002;**90**(9):863–9.
48. Elshal AM, Mekkawy R, Laymon M, et al. Holmium laser enucleation of the prostate for treatment for large-sized benign prostate hyperplasia; is it a realistic endourologic alternative in developing country? *World J Urol* March 2016;**34**(3):399–405.

Thulium Laser Prostatectomy

12

Henry Tran[1,2], *Doreen E. Chung*[1]
[1]Columbia University Medical Center/New York Presbyterian Hospital, New York, NY, United States; [2]University of British Columbia, Vancouver, Canada

Introduction

The thulium laser was first introduced in 2005 and has several unique properties making it useful in urology. The peak absorption of water in tissue is 1.94 μm and the laser has a wavelength range of 1.75–2.22 μm. Because the wavelength is very close to the peak absorption of water, more efficient vaporization with shallower depth of penetration can be achieved.[8] The laser fiber can operate in either a pulsed or continuous wave mode because it is diode pumped. Typically, a pulsed mode is excellent for treating urolithiasis while a continuous mode is better for hemostasis and tissue coagulation, giving the laser useful applications in the kidney, ureter, and bladder.[7,9–11] The thulium laser has properties making it ideal and comparable to the holmium and 532 nm (GreenLight) lasers. In this chapter, we will provide an overview of the various thulium laser surgical techniques for the treatment of benign prostatic hyperplasia (BPH). We will also outline current studies on these techniques including studies comparing thulium laser prostatectomy to other minimally invasive techniques for BPH.

Laser Properties

Prior to 2005, the thulium laser used in medicine had a maximum power output of 5 W, limiting its usefulness for urologic applications. In 2005, a new system increased power output to 40 W, expanding the possible applications to urology.[8] The thulium laser has several unique properties that distinguish it from the other two lasers commonly used for the treatment of BPH, the holmium laser and the 532 nm (GreenLight) laser. The holmium laser has a wavelength 2140 nm, with a depth of penetration 0.5 mm.[12] At 2010 nm, the thulium laser has a similar wavelength that is closer to the peak absorption in water (1940 nm). Thulium laser energy is delivered in a continuous wave, whereas the holmium laser is delivered in a pulsed wave. These properties of the thulium laser allow for a shallower depth of penetration in tissue (0.2 mm vs. 0.4 mm) and a fourfold decrease in the zone of thermal damage compared to the holmium laser.[1,13] Whereas postoperative dysuria is a common adverse effect seen with the GreenLight laser, the smaller zone of thermal damage with the thulium may decrease this side effect. In comparison, the GreenLight has a depth of penetration of 0.8 mm and coagulation zone of 1–2 mm.[12] Standard silica optical fibers in a range

A Comprehensive Guide to the Prostate. https://doi.org/10.1016/B978-0-12-811464-3.00012-X

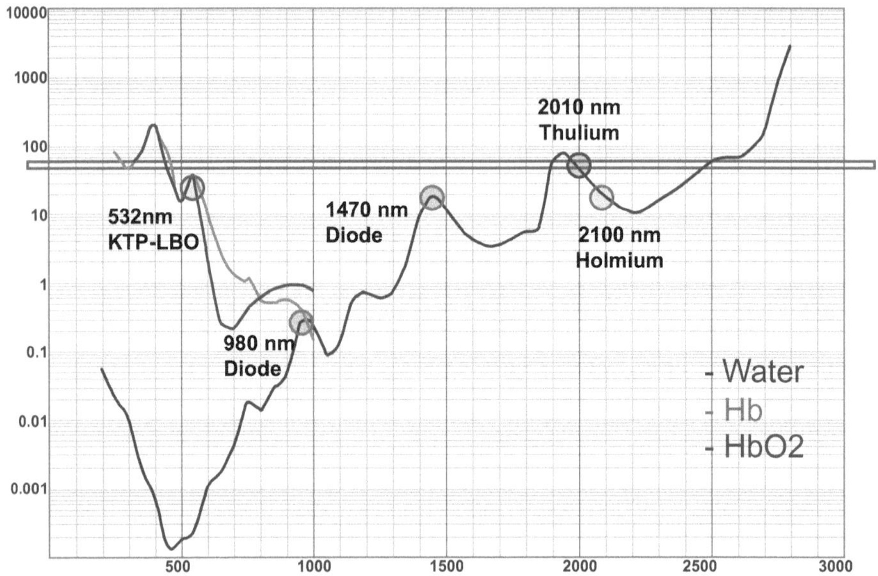

Figure 12.1 Peak absorption spectra for various laser energy types.

of diameters can house the thulium laser beam. Silica fibers are robust, reusable, and flexible, which increases usability and decreases cost.[14] Fig. 12.1 outlines the peak absorption spectra for various commonly used laser types.

Ex-Vivo Studies

Fried et al. conducted studies to assess the effects of a 40-W thulium laser (Model TLR-40-1940; IPG Photonics, Oxford, MA) on canine prostates, canine bladders, and porcine ureters.[8] This study showed good promise for use of the thulium laser for prostate vaporization and bladder neck contracture incision. To demonstrate utility of the laser for prostate vaporization, eight canine prostates with a mean mass 16.4 ± 7.2 g were vaporized and the efficacy rate was 7.6 kJ/g. The 40-W thulium laser ablation rate shown in this study is lower compared to the holmium laser and the 80 and 100-W GreenLight lasers, which typically report ablation rates of 1–2 g/min.[15,16] However, higher ablation rates are currently possible with the newer higher power thulium systems. The larger zone of thermal coagulation defect seen in this study was likely due to the longer pulse-width of 10 ms used, and higher power lasers should be able to provide adequate tissue ablation with shorter pulse-widths.

Another ex-vivo study used porcine kidneys to compare the tissue ablation rates between standard transurethral resection of prostate (TURP), 80-W potassium-titanyl-phosphate (KTP; GreenLight, Laserscope, San Jose, CA) laser, and a 70-W thulium laser. The authors found that TURP had the highest ablation rate—resecting 8.3 ± 0.4 g in 30 s. The thulium had a higher ablation rate (6.6 ± 0.7 g after 10 min) compared to

KTP (4 ± 0.5 g after 10 min).[2] However, this study did not evaluate the tissue ablation rate of the more contemporary higher power GreenLight XPS. One notable advantage of the lasers compared to TURP seen in the study was the bleeding rate. The bleeding rate with TURP was 20.14 ± 2.03 g/min compared to that of the thulium laser (0.16 ± 0.07 g/min) and KTP laser (0.21 ± 0.07 g/min). The KTP laser had the deepest coagulation zone depth (667 ± 64 μm), while the thulium laser (265 ± 41 μm) and TURP (287 ± 28 μm) were similar.

Several surgical techniques using the thulium laser have been described for laser prostatectomy including thulium vapoenucleation (ThuVEP), thulium laser resection of the prostate-tangerine technique (TmLRP), thulium laser enucleation (ThuLEP), and thulium vaporization (ThuVP).

Thulium Vapoenucleation

Vapoenucleation is the most commonly described technique and involves trilobar enucleation using a combination of laser and mechanical energy followed by intravesical morcellation, similar to holmium laser enucleation of prostate (HoLEP).[1,2] The thulium laser can perform simultaneous vaporization and resection, which is the reason the technique has been coined "vapoenucleation." Furthermore the apical tissue is usually vaporized as opposed to enucleated to minimize damage to the external sphincter.[17] The majority of studies are short-term and there is large heterogeneity in patient populations making the studies difficult to compare.

Initial human studies were done on small prostates, using the 70-W RevoLix (Lisa Laser Products, Katlenburg-Lindau, Germany) laser system in a small number of patients.[17,18] ThuVEP, enucleation, and laser resection studies are shown in Table 12.1. Bach et al. published a single surgeon feasibility study of the procedure performed on 54 men with small prostate volume (mean 30.3 [range 12–38] grams).[17]

Despite small prostate size, 26% of patients were in retention prior to surgery. Mean resection time was 1.5 g/min. The average hospital stay was 3.5 (2–6) days, with a mean catheter time of 1.7 (1–3) days. No patients required transfusion or rehospitalization. There were improvements seen in International Prostate Symptom Score (IPSS) (19.8–6.9), Qmax (4.1–20.1 mL/s), and postvoid residual volume (PVR) (86–12 mL).

Following this initial study, this same group of authors investigated the short-term outcomes (follow-up until discharge) of thulium:YAG (Tm:YAG) laser vapoenucleation of the prostate in patients with larger prostates with the 70-W RevoLix Tm:YAG laser.[19] 88 patients with prostate volume 61.3 ± 24.0 mL underwent ThuVEP with the 70-W RevoLix Tm:YAG laser. Morcellation was performed in all patients following enucleation, which differed from the prior feasibility study.

The study inclusion criteria were patients with Qmax <15 mL/s and IPSS >7. All patients underwent continuous bladder irrigation overnight after surgery.

Mean age was 71 ± 8 years. Of these patients, 31% were being treated for acute urinary retention. Mean operative time was 72 ± 27 min. Mean laser energy per case was 124 ± 40 kJ with 32 ± 18 g of tissue obtained. Because the thulium laser vaporizes

Table 12.1 Thulium vapoenucleation and laser resection studies[40–42]

Study	No. subjects	Mean/median prostate size (cc's)	OR time (min)	Laser time (min)	Follow-up (months)	Complications
Bach et al.[17]	54	30 (12–38)		52 (28–72)	12	11% UTI
Xia et al.[13]	52	59 ± 17	46 ± 16	Not reported	12	4% UTI 0% transfusion
Bach et al.[19]	88	61 ± 24	72 ± 27	32 ± 10	n/a	17% overall 2% transfusion 7% UTI 2% reoperation
Bach et al.[20]	208	46 ± 22 and 43 ± 24	72 ± 29 and 68 ± 29	36 ± 12 and 34 ± 12	n/a	1% transfusion 8% UTI 3% reoperation
Gross et al.[3]	1080	51 (IQR 36–78)	56 (IQR 40–80)	33 (IQR 33–71)	n/a	2% UTI 2% transfusion 5% reoperation
Zhang et al.[38]	71	76 ± 10	72 ± 19		18	0% transfusion
Macchione et al.[5]	76	65 ± 15	47 ± 8	Not reported	6	7% UTI 1% transfusion 0% reoperation
Peng et al.[36]	50	58 ± 12	61 ± 24	Not reported	3	0% transfusion
Pearce et al.[6]	25	163 ± 62	204 ± 58	63 ± 20	12	36% overall 20% UTI 12% retention 4% transfusion

and enucleates tissue, the amount of tissue retrieved is not directly comparable to that of the HoLEP procedure. Four (5%) patients were found to have incidental prostate adenocarcinoma. Complications were seen in 16.6% of patients. Two (2.2%) patients required blood transfusions for bleeding, three (3.4%) patients did not require transfusions but also had intra or postoperative bleeding, six (6.8%) patients developed urinary tract infection (UTIs), and two (2.2%) patients were unable to void and underwent an adjuvant procedure immediately. Qmax (3.5 ± 4.7–19.8 ± 11.6 mL/s, $P < .001$) and PVR (121 ± 340–22 ± 33 mL, $P = .03$) both improved significantly immediately after surgery.

Bach et al. subsequently compared outcomes of ThuVEP using the 70-W RevoLix laser system in patients with recurrent urinary retention with and without indwelling Foley catheters.[20] 65 patients with recurrent retention before surgery were compared to 143 patients without retention or indwelling Foley catheter. These patients had similar prostate volumes (46 ± 22 and 43 ± 24 mL, $P = .18$). Follow-up was again only recorded until 1 day after discharge. Transfusion rate overall was only 1%. Ten (15.4%) patients in the urinary retention group developed a UTI compared to six (4.2%) in the nonretention group. ThuVEP appeared safe and effective in the short term even for patients in retention.

This same group of authors reported on a large series of 1080 patients examining immediate postoperative outcomes and complications of ThuVEP by 11 surgeons at a single institution.[3] The study used a 2-μm continuous-wave thulium Tm:YAG laser (RevoLix; LISA) as the energy source at 70, 90, 120, and 200 W. Median age was 71 years (IQR: 66–76 years), and average prostate size was 52 mL (IQR: 36–78.7 mL). Median resected tissue weight was 30 g (IQR: 16–51.25 g). Following surgery, improvements were seen in Qmax (8.9 mL/s to 18.4 mL/s, $P < .001$) and PVR (120 mL–20 mL, $P < .001$). There were no mortalities and the overall morbidity rate was 16.9%. Minor complications (Clavien I and II) occurred in 24.6% of patients. 6.6% of patients required intervention within 4 weeks (Clavien III). One patient developed a myocardial infarction (Clavien IVa). The most common postoperative complications included urinary retention (9%), UTI (6.9%), clot retention without surgical intervention (3.5%), residual prostate tissue requiring reoperation (2.7%), capsular perforation (2.1%), hemorrhage requiring reoperation (2%), hemorrhage not requiring reoperation but requiring blood transfusion (1.7%), UTI with bacteremia (1.5%), extraperitoneal fluid collection (1.5%), superficial bladder injury due to morcellation (1.4%), ureteral orifice injury without hydronephrosis (0.7%), ureteral orifice injury with hydronephrosis (0.6%). The readmission rate was 4.1% and the reoperation rate was 4.7%. A subanalysis stratified patients by prostate size (<40, 40–79, and ≥80 mL) found no difference in complication rates. The observed immediate outcomes and complication rates were comparable to similar large series for HoLEP and Photovaporization of the prostate (PVP) and appeared lower than those seen for TURP and open simple prostatectomy. This large series confirmed that ThuVEP is a safe and effective procedure for the treatment of symptomatic BPH with low perioperative morbidity.

Netsch et al. compared the 120-W to the 70-W thulium laser and found that the higher power laser had higher ablation with relatively comparable tissue penetration depth and rate of bleeding.[21] Macchione et al. then used the 120-W thulium laser and performed ThuVEP in 76 patients who were anticoagulated.[5] 41 patients with mean prostate size 65 ± 15 mL underwent surgery while continuing oral anticoagulation and their outcomes were compared to 35 patients who had their anticoagulation held 10 days before surgery with bridging to low molecular weight heparin. One single anticoagulated patient required a transfusion. 4% of patients required CBI for hematuria among all groups, and both groups had significantly improved Qmax, IPSS, PVR, and QoL at 3 and 6 months following surgery. This study demonstrated that ThuVEP can be used safely and effectively in patients on active oral anticoagulation. There are many strengths of this large series of patients including adequate power, multiple surgeons, and use of standardized reporting methods for complications. The fact that it was a single center study and that various generations of the thulium laser were used may not make the results generalizable to some populations.

Pearce et al. recently looked at outcomes up to 12 months postoperatively in 25 patients from North America with large prostates (>75 mL) who underwent ThuVEP using a different thulium laser system, the 150-W CyberTM (Quanta System, Italy). This is the only ThuVEP series to come out of North America. Mean age was 70 ± 9 years, and mean prostate size was 163 ± 62 g. Mean age-adjusted Charlson comorbidity index was 4.2 ± 2 and mean ASA class was 2.6 ± 0.6. Prior to surgery, 84% of patients were in urinary retention, and 40% were taking oral anticoagulation. Mean hospital stay was much shorter than in previous studies, at 1.2 ± 1.2 days, and 28% of patients were discharged on the same day as surgery. Improvements were seen compared to baseline at 3, 6, and 12 months in IPSS (19.3 ± 7.3 vs. 5.8 ± 4.1, 5.4 ± 5.6, 4.7 ± 4.6, $P < .05$), QoL (5.2 ± 1.3 vs. 2.3 ± 2, 1.2 ± 1, 1.2 ± 1.3, $P < .05$), Qmax (mL/s) (4.2 ± 4.2 vs. 17.7 ± 7.2, 15.0 ± 9.9, 20.1 ± 10.4, $P < .05$), and PVR (mL) (355 ± 274 vs. 67 ± 93, 102 ± 175, 50 ± 58). There were 9 Clavien grade I complications, 1 grade II complication, and no grade ≥ III complications. Grade I complications included 5 culture-proven UTIs (20%) treated with oral antimicrobial agents, 3 patients failed initial voiding trial (12%) requiring recatheterization, and 1 patient (4%) experienced gross hematuria with clot retention requiring bladder irrigation. The single grade II complication was a blood transfusion for a patient who had significant cardiac comorbidity and was anemic prior to surgery. This series further supported that ThuVEP is a very safe and effective treatment even in patients with large prostate gland size and high comorbidity up to 12 months.[6]

Like all surgical procedures, there is a learning curve for ThuVEP. Gross et al. divided their 1080 patients into five consecutive subgroups of 216 patients each, and overall complication rates decreased over time. The first subgroup had a complication rate of 41.7% compared to the last subgroup with a complication rate of 19.4%. Transfusion rates improved from 3.2% to 0%. Clavien IIIb complications also decreased between the first group (11.1%) and last group (3.3%).[3]

Netsch et al. studied prospectively the learning curve of ThuVEP and found that among three surgeons with different levels of experience with the operation, there were differences in enucleation, operative time, and amount of laser energy used; however, Qmax, PVR, and IPSS improvements were seen in patients from all groups and no significant difference in complication rates were seen.[22] These studies by the same group are conflicting but suggest that while there is a learning curve, it can be overcome within a reasonable number of cases. While HoLEP studies support the fact that enucleation takes many cases to become proficient at, fewer cases in ThuVEP should be required since there is a vaporization component and vaporization is a skill that the majority of urologists are comfortable with.

Thulium Laser Enucleation of the Prostate

Thulium laser enucleation of the prostate (ThuLEP) is extremely similar to ThuVEP with the only difference being that enucleation of the adenoma is done mechanically without the use of energy. The initial cuts are still made with the laser.[3] Due to the continuous pulse, whenever the laser is employed vaporization occurs. In papers, the description between the two techniques is often indistinguishable.[2]

Carmignani et al. showed that ThuLEP, like ThuVEP, is safe and effective even in patients with a history of refractory urinary retention and indwelling Foley catheters.[23] 93 patients were found eligible for the study and all successfully voided within 7 days following surgery. Average hospital stay was 2.3 ± 1.7 days and no patients needed further catheterization at 90 days. Significant improvements were seen in IPSS, Qmax, and PVR at 7, 30, and 90 days following surgery. The average hospital stay was 2.3 ± 1.7 days. None of the patients required Foley catheter at 90 days, and all reported complications occurred within 30 days. There were 11 patients with Clavien grade I and 1 patient with Clavien grade II complications. Two patients had a PVR >100 mL and four patients had transient urgency or incontinence, but all cases resolved within 4 weeks.

Iacono et al. reported on a series of 148 men that underwent ThuLEP with large prostate size (108 ± 24 mL) using the 120-W RevoLix laser with 12 months of follow-up. Significant improvements were seen in IPSS, QoL, and PVR. Complications included UTI in 19 patients (12.8%), postoperative urgency incontinence in 10 (6.7%), early recatheterization with residual tissue at prostatic apex in 4 patients (2.7%), blood transfusion in 4 patients (2.7%), and bladder injury during morcellation in 2 patients (1.3%). There was significant increase in IIEF but the absolute difference in score was very small (19.3 ± 8.2–20.3 ± 8.2, $P < .05$).[24]

Rausch et al. performed ThuLEP in 234 patients with mean prostate size of 84.8 ± 34.9 mL, from 2008 to 2012, using the RevoLix (power not specified). The 30-day complication rate was 19.7%. In the perioperative period only 3% of patients experienced UTI, 6.8% required catheter replacement, and 0.9% required blood transfusion. Following surgery 3% experienced urgency incontinence and 2.1% experienced bladder neck contracture.[4] Complications occurring after 30 days included stress incontinence in one patient (0.5%) and inability to void requiring chronic catheterization in 19 (8.1%) patients. Predictors of complications in univariate and multivariate analysis were aged >80 years and prostate size <50 mL.[25]

Few studies have addressed sexual outcomes in following thulium laser prostatectomy. Carmignani et al. evaluated sexual outcomes in 110 men undergoing ThuLEP with the Cyber 150-W laser in a prospective study using validated instruments. Patients were evaluated before surgery and at 3 and 6 months after ThuLEP with IPSS, IIEF-5, and ICIQ-MLUTSsex.[26] While a significant and sustained improvement in the scores evaluating urinary symptoms were observed, no significant differences were seen in erectile function before and after surgery.[26]

Thulium Laser Resection of the Prostate-Tangerine Technique

The tangerine technique is similar to ThuVEP; however, this technique is described as being similar to peeling a tangerine. An incision is made at the 5 and 7 o'clock until the surgical capsule is reached from bladder neck to the verumontanum. The median lobe is then removed via a transverse section and the lateral lobes are then resected by cutting in a large curve after making the transverse incision. The prostate tissue is dissected off of the surgical capsule and resected into small pieces using a combination of semicircular and transverse incisions.

Zhuo et al. demonstrated that TmLRP is an effective technique for treating patients presenting with acute urinary retention from their BPH.[27] 52 patients with mean age 70.3 (54–89 years) were evaluated using IPSS, QoL, PSA, transrectral ultrasound (TRUS), PVR, and Qmax. All patients who required a Foley catheter prior to surgery were able to void immediately and 6 months after surgery. Significant improvements ($P<.001$) in mean IPSS (21.6 ± 6.8–6.8 ± 3.2), QoL (4.4 ± 1.2–0.9 ± 0.8), Qmax (0–18.7 ± 6.9 mL/s), and PVR (274.5 ± 150.7 to 40.6 ± 22.5 mL) were seen at 6 months. Regarding adverse events, no patients had TUR syndrome, two patients (3.8%) needed transfusion, five (9.6%) had transient hematuria, two (3.8%) needed 3-day catheterization for clot urinary retention, three (5.8%) had UTI requiring antibiotics, and one (1.9%) had prostate capsule perforation. Seven patients (14%) had some urinary incontinence within the first month after surgery, but this resolved in all patients within 30 days. In the sexually active patients retrograde ejaculation was seen in 18 patients (35%), and four patients developed a urethral stricture requiring subsequent urethrotomy. This initial study suggested that TmLRP is effective and safe even for patients in urinary retention.[27]

The same group of authors reported on a series of 95 consecutive patients who underwent TmLRP with large prostate size (>80 mL). Mean prostate volume was 106.81 ± 24.79 mL and mean resected weight of tissue was 64.08 ± 12.89 g. Catheterization time was 2.25 ± 0.9 days and mean hospitalization was 5.39 ± 1.18 days. Follow-up occurred up to 18 months following surgery. For the time period up to 18 months following surgery there were significant improvements in IPSS (20.01 ± 7.08 vs. 4.96 ± 3.68), QoL (4.10 ± 1.16 vs. 1.23 ± 1.30), Q_{max} (8.14 ± 3.81 mL/s vs. 18.33 ± 2.56 mL/s), and PVR (102.70 ± 70.64 mL vs. 20.28 ± 30.02 mL), compared with baseline values ($P<.001$). IIEF-5 remained stable. Complication rate was 10.5%. There were no Clavien ≥ III complications.[28]

A recent large, multicenter study by Sun et al. evaluated 2216 patients who underwent TmLRP for BPH at four centers with long-term follow-up data.[29] Of the 2216 patients, 1353 (61.1%), 1129 (50.9%), 847 (38.2%), and 541 (24.4%) patients were followed at 5, 6, 7, and 8 years respectively. The incidence of minor Clavien I complications was 21.5% and included UTI, urinary retention, urinary incontinence, capsular perforation, and clot retention. Clavien II complications were seen in 2.3% of patients, including thrombosis (0.4%), severe infection requiring IV antibiotics (1.3%), and severe bleeding with transfusion (0.6%). Major complications (Clavien III) requiring repeat intervention occurred in 2.2% of patients including residual prostate tissue requiring re-resection (0.5%), severe bleeding requiring surgical intervention (0.2%), and clot urinary retention requiring cystoscopy (1.5%). No Clavien IV or V complications were found. Long-term complications included urethral strictures in 58 (2.6%) patients and bladder neck contractures in 35 (1.6%) patients. The majority of these long-term complications occurred within the first year after surgery. Other long-term issues included BPH recurrence requiring reoperation in 27 cases (1.2%) within 5 years and persistent stress urinary incontinence (SUI) amenable to conservative treatment in 2 (0.1%) patients. Durable improvements in various outcome parameters were seen at 3 months, 1 year, and 8 years respectively for median IPSS (7, 6, 6), median QoL (1, 1, 1), median Qmax (21.7, 22.3, 19.2 mL/s), and PVR (22, 18, 20 mL).[29]

Overall, like the other enucleation techniques, TmLRP is a highly effective technique and can be used in patients presenting with urinary retention, and large prostate glands. Durable improvements in objective and subjective parameters have been demonstrated over the long term.

Thulium Laser Vaporization of the Prostate

Although not as well studied, another highly effective technique for thulium laser prostatectomy is vaporization (ThuVP). Fig. 12.2 outlines the technique and steps in ThuVP. Pariser et al., in 2014 reported on short-term outcomes of ThuVP using the 150-W Cyber thulium laser in 59 patients with mean prostate volume 57 ± 30.2 mL. This was the first study to be published on outcomes on any technique for of thulium laser prostatectomy in a North American patient population. 47% of patients were taking oral anticoagulation and 78% were discharged home on same day of surgery. Significant improvements in IPSS (8.7 ± 6.5 vs. 19.9 ± 8, $P < .05$) and QoL (2.1 ± 1.4 vs. 4.5 ± 1.1, $P < .05$) were seen at 3 months. Significant improvements were also seen at 1 week, 1 month, and 3 months in Qmax and PVR. No blood transfusions were administered and there were no Clavien grade ≥ III complications within 30 days of surgery. The 30-day complication rate was 20% with all complications being Clavien grade I and II. 10% required catheterization for urinary retention and 10% developed symptomatic culture-proven UTIs. Between 30 and 90 days, one patient developed a bladder neck contracture and one developed a urethral stricture. Both were managed endoscopically.[30] All 15 patients who presented in retention were voiding spontaneously at the time of their last follow-up. This study demonstrated that ThuVP is safe and effective even in patients who are in retention and in those taking oral anticoagulation.

Around the same time, Vargas et al. also published outcomes of the 150-W ThuVP with 6 months of follow-up in 55 patients out of Spain.[31] Prostate size was comparatively smaller with median 51 mL (IQR 36–78.7 mL) and patients on anticoagulation were excluded. As in Pearce et al.'s study, patients were also discharged home the

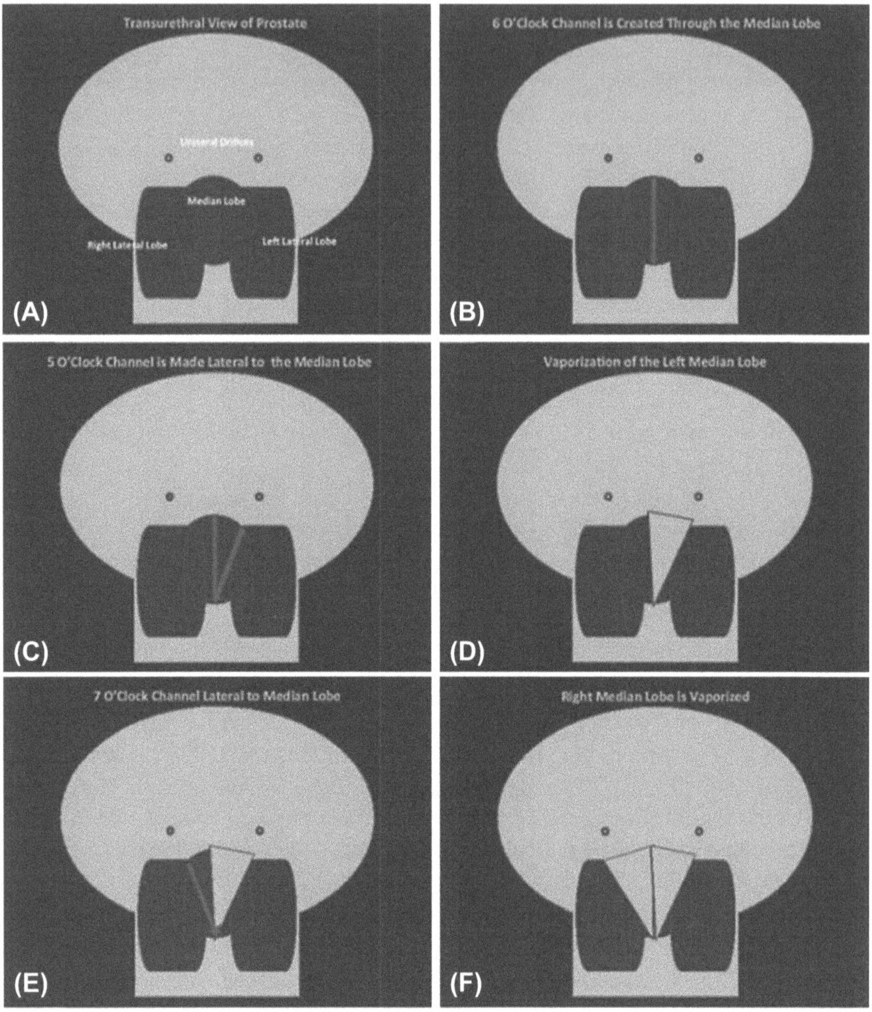

Figure 12.2 (A) Transurethral view of a prostate with trilobar hypertrophy. (B) Groove is created at the 6 o'clock position to facilitate irrigation flow and visualization of ureteral orifices. (C) A channel is created at the 5 o'clock position lateral to the median lobe to separate the median from the lateral lobes. (D) Vaporization of left median lobe between these two channels is then performed. (E) A channel is then created at the 7 o'clock position to separate the right median and lateral lobes. (F) The right median lobe is vaporized. (G) Next, a groove is created at the 2 o'clock position, dropping down the left lateral lobe. (H) The left lateral lobe is vaporized. (I) 10 o'clock groove is created dropping down the right lateral lobe. (J) The right lateral lobe is vaporized. (K) The anterior tissue between 10 o'clock and 2 o'clock is vaporized. Remaining apical tissue is also vaporized.

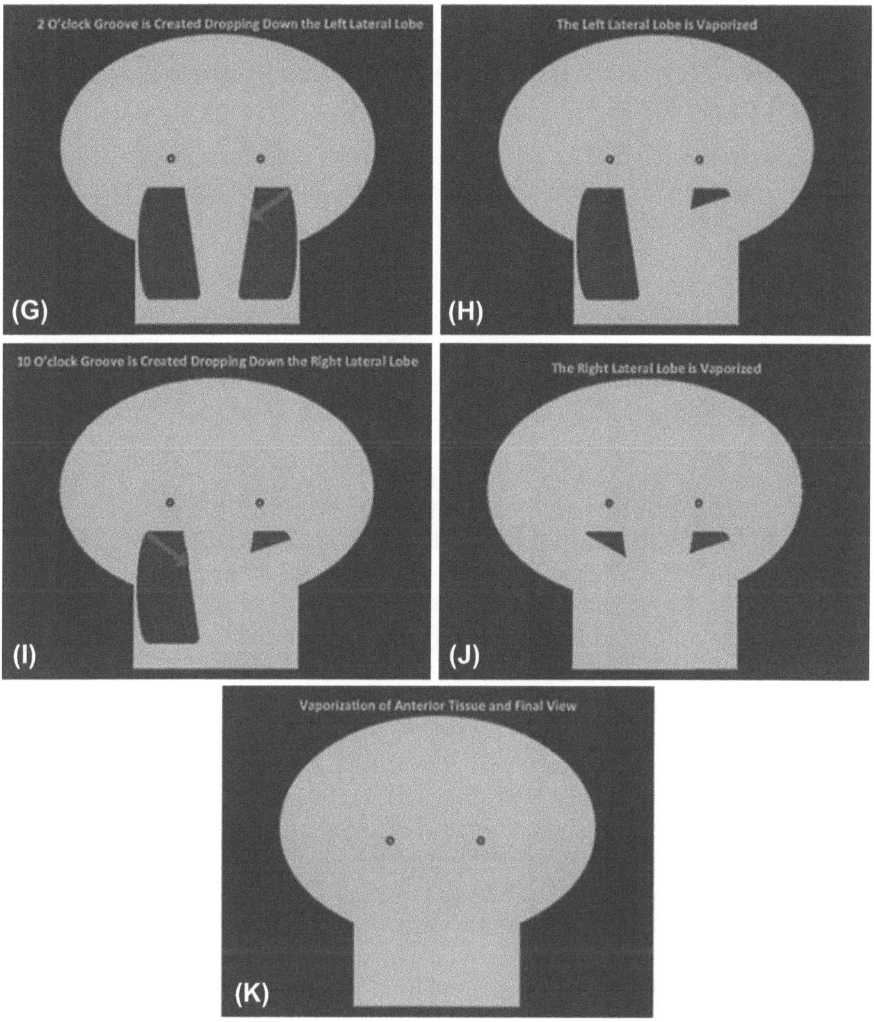

Figure 12.2 *(Continued)*

same day of surgery. Significant improvements were seen in Qmax (mean improvement 9.33 mL/s) and mean IPSS (reduction by 17 points). Immediate complications were recorded at 1 month follow-up and included acute urinary retention (one patient), urinary tract infection (two patients), and gross hematuria (two patients). The only late complication observed was a single patient with bladder neck contracture at 6 months follow-up. There were no Clavien grade III or higher complications within 30 days.[6] No comparative data are available for ThuVP; however, this technique appears to have short-term complications similar to those described after vaporization with other laser systems as well as TURP.[7,8]

To date, early short-term studies on ThuVP appear promising and outcomes appear similar to those from PVP but longer-term comparative studies are needed.

Comparative Studies

Comparative Studies to Traditional Monopolar Transurethral Resection of Prostate

The gold standard for surgical treatment of BPH has been monopolar TURP. There are a number of studies that compare thulium laser prostatectomy with this operation.[18] Xia et al. used a 50-W continuous wave thulium laser (LISA laser products OHG, Germany) to perform prostate resection using the tangerine technique (TmLRP) and compared their 12 month outcomes to standard TURP.[18] Patients with prostate volume <100 g and urodynamic evidence of outlet obstruction were included in the study. Those in retention were excluded. 48 patients were randomized to TURP and 52 patients to TmLRP.

The mean age of patients in both groups was 69 ± 7 years. Mean prostate volume via TRUS (59 ± 17 mL and 55 ± 16 mL), Qmax (8 ± 3 and 8 ± 3 mL/s), PVR (93 ± 32 and 85 ± 37 mL), and baseline IPSS (22 ± 7, 21 ± 6) was similar for the TmLRP and TURP groups. Operative times (TmLRP 46 ± 16 and TURP 50 ± 20 min) were similar. The resected weight for TmLRP was lower than the TURP group (19 ± 14 and 21 ± 10 g, $P = .0001$). In the TmLRP group there was less decrease in hemoglobin compared to the TURP group (0.92 ± 0.82 vs. 1.46 ± 0.65 g/dL, $P = .0004$). Hospital stay (115 ± 24 vs. 161 ± 34 h, $P < .0001$) and catheterization time (46 ± 26 and 87 ± 34 h, $P < .0001$) were also shorter with TmLRP arm. No differences between the procedures in IPSS, QoL, Qmax, and PVR were found at 1, 6, and 12 months. Transfusion rate was 4% (n = 2) in the TURP group versus 0% in the TmLRP group. No significant differences were found in rates of transfusion, UTI, TUR syndrome, de novo SUI, or urethral stricture development. This study suggested that TmLRP compared to TURP is associated with lower morbidity but similar subjective and objective patient improvements.

Cui et al. performed a randomized trial comparing safety and efficacy of monopolar TURP compared to TmLRP with intermediate-term outcomes to 4 years. 47 and 49 patients were randomized to the TmLRP and TURP groups. These authors found that the degree of improvement in all micturition parameters was comparable in both the TmLRP and the TURP patients at every time point of follow-up to 48 months including IPSS, QoL, PVR, and Qmax. Details of the short-term complications were not reported. Long-term reoperation rates for BPH recurrence and rate of bladder neck contracture were similar in the two groups, although the study may be underpowered to look at these parameters.[32]

Chang et al. recently reported a study comparing high-power 150-W CyberTM ThuVEP to standard TURP with 1 year follow-up.[33] 29 patients were assigned to have ThuVEP and 30 were assigned to TURP. Voided volume, Qmax, PVR, and mean flow rate all improved significantly in both groups and no differences were seen in IPSS or QoL after 1 year of follow-up. Transfusion rate in this study was lower for the ThuVEP group at 13.8% versus 26.7% in the TURP group. This transfusion rate was higher than that reported in other studies using the high power laser.[3,30] However, there was minimal change in hemoglobin level before and after the operation (ThuVEP: 0.5 ± 1.3 g/dL, TURP: 0.5 ± 1.1 g/dL).[33]

A metaanalysis including nine trials was done to examine the performance of TmLRP versus TURP. TmLRP was found to be a safe, feasible, and efficient alternative to TURP for treating patients with BPH with reliable perioperative safety, fewer complications, and comparable efficacy in relation to Qmax, PVP, QoL, and IPSS. Patients undergoing TmLRP experienced smaller declines in serum sodium levels ($P<.0001$), hemoglobin levels ($P<.0001$), shorter durations of catheterization ($P<.0001$), shorter lengths of hospital stay ($P<.0001$), and fewer total complications ($P<.0001$).[34]

A large systematic review and metaanalysis by Zhu et al., in 2015 was done comparing TmLRP and TURP.[35] Seven trials with four randomized controlled trials (RCTs) were assessed. The TmLRP procedure had a longer mean weighted operative time by 8.18 min, but TmLRP had significantly less sodium decrease (WMD = −3.73 mmol/L, 95% CI −4.41 to −3.05, $P<.001$), shorter catheterization time (WMD = −1.29 days, 95% CI −1.81 to −1.29, $P<.001$), and less transfusion (OR 0.09; 95% CI 0.02–0.41, $P=.002$). At 1, 3, and 12 months follow-up, there were no difference in IPSS, QoL, Qmax, and PVR. Overall, TmLRP had similar efficacy with less morbidity in terms of blood transfusions, serum sodium decrease, catheterization time, and hospital stay but TURP had shorter operative times.[35]

Multiple comparative studies between thulium laser prostatectomy and monopolar TURP suggest that thulium laser prostatectomy appears to have similar outcomes in terms of IPSS, QoL, Qmax, and PVR with better morbidity outcomes such as blood transfusion rate, serum sodium decrease, hospitalization length.

Comparative Studies to Bipolar Transurethral Plasmakinetic Prostatectomy

Peng et al. compared short-term outcomes of TmLRP using the 50-W thulium laser (LISA Laser products OHG, Germany) to bipolar transurethral prostatectomy (BiTURP).[36] 100 patients were randomized to either treatment. Baseline characteristics appeared similar between the TmLRP and BiTURP patients for mean prostate volume (58 ± 12 and 58 ± 15 mL), IPSS (20 ± 8 and 19 ± 8), QoL score (5 ± 3 and 5 ± 1), Qmax (8 ± 4 mL/s for both groups), and PVR (97 ± 35 and 88 ± 38 mL). TmLRP was associated with shorter catheterization time (1.8 ± 0.4 vs. 3.2 ± 0.6 days, $P<.05$) and hospital stay (3.3 ± 0.8 vs. 4.1 ± 1.3 days, $P<.05$) compared to BiTURP with a trade-off of longer operative time (61 ± 24 vs. 30 ± 6 min, $P<.05$). Postoperative serum sodium and hemoglobin levels were similar. No transfusions were necessary in neither group. Significant improvements were seen in both groups in subjective and objective outcomes at 1 and 3 months postop. Short-term and perioperative complications were not mentioned. Overall, TmLRP appeared safer, with less blood loss and urethral stricture rate and similar improvements in IPSS, Qmax, and PVR when compared to BiTURP.

Another study done by Yang et al., evaluated 158 patients randomized to 100-W thulium laser TmLRP versus BiTURP with 18-months follow-up.[37] Baseline parameters for TmLRP and BiTURP—prostate volume (62 ± 7 and 69 ± 23 mL), IPSS (23 and 23), and age (62 ± 7 years and 61 ± 7 years)—were similar. As in the previous study, TmLRP had a longer operative time (65.4 ± 22.2 min vs. 47.4 ± 15.9 min, $P=.022$) but also had shorter hospital stay (2.5 ± 1.4 vs. 4.6 ± 1.4 days, $P=.026$), shorter catheterization time (2.1 ± 0.8 vs. 3.5 ± 1.2 days, $P=.031$), and less decrease in hemoglobin (−0.15 ± 0.02 vs. −0.30 ± 0.03 g/dL, $P=.045$). Similar improvements were seen in

IPSS, QoL, Qmax, and PVR. Short-term and perioperative complications were not mentioned. No patients in any group developed urethral strictures or bladder neck contractures. No transfusions were necessary in either group and clot retention was seen in 4% of BiTURP and 1% of TmLRP patients.[37]

Wei et al. compared the efficacy and safety of the TmLRP using the 120-W Tm:YAG laser to BiTURP in patients with prostate size >80 g.[4] 45 patients were randomized to each group and were evaluated at 1, 6, 12, and 18 months. The mean prostate sizes (112.86 ± 28.36 mL TmLRP, 115 ± 39.45 mL BiTURP group) as well as other perioperative parameters were similar in both groups. Compared to the BiTURP group, the TmLRP group had statistically less hemoglobin decrease, shorter catheterization time, and hospital stay. Both groups had significant postoperative improvements in IPSS, QoL, Qmax, and PVR, with no differences observed between the two groups. Both techniques showed a low perioperative complication rate (TmLRP vs. BiTURP: Clavien I: 13.33 vs. 20%; Clavien II: 0 vs. 2.22%; Clavien III: 0 vs. 0%). One patient in the BiTURP group required a blood transfusion perioperatively. Urethral stricture rate was equal (one patient in each group), and one patient in the BiTURP group developed a bladder neck contracture. Additionally, there were no significant reduction in IIEF-5 scores during the follow-up period compared with preoperative data, and no statistically significant difference was observed between the two groups.[4]

Comparative Studies to Holmium Laser Enucleation

Comparison studies between ThuLEP and holmium laser enucleation (HoLEP) have been performed. In both techniques, the median and lateral lobes are demarcated and lifted up off the capsule. Next, the lobes are then left attached to the bladder neck. The main difference for ThuLEP though is that the enucleation of the adenoma is done mechanically, without any laser energy.[1] Zhang et al. used the 70-W (Revolix, Katlenburg, Germany) thulium laser system and the 100-W (VersaPulse, Lumenis, Santa Clara, CA) holmium laser system, and randomized 133 patients to ThuLEP and HoLEP.[38] A monopolar electrocautery loop is then used to mechanically fragment the lobes into small pieces. ThuLEP was performed in 71 patients and HoLEP was performed in 62 patients. Baseline characteristics were similar between ThuLEP and HoLEP for age (76 ± 10 and 74 ± 10 years), and mean prostate volume (47 ± 25 and 44 ± 23). Operative time was longer for ThuLEP compared to HoLEP (72 ± 19 vs. 62 ± 20 min); however, weight of resected tissue (38 ± 12 vs. 40 ± 10 g, $P = .072$), decrease in hemoglobin (0.5 ± 0.1 vs. 0.5 ± 0.1 g/dL, $P = .107$), and catheterization time (2.4 ± 10 vs. 2.5 ± 1.0 days) were comparable. Outcomes for IPSS, QoL, Qmax, PVR, and PSA reduction were similar up to and including 18 months. No transfusions were needed for either group. One patient in each of the ThuLEP and HoLEP groups required suprapubic catheter insertion for intraoperative capsular perforation.[38] No transfusions were required in either group. No short- or long-term reoperations were mentioned.[38]

The numerous studies in the literature show that enucleation of the prostate with both thulium and HoLEP lasers is comparably safe and effective, with similarly good clinical outcomes compared to standard TURP. These methods can be effective even in patients with urinary retention.[39]

Conclusions

In summary, the thulium laser has several unique properties that make it a safe and effective laser for treatment of BPH. The thulium laser is highly versatile and can be utilized in various ways to perform laser prostatectomy including vaporization, enucleation, and vapoenucleation. ThuVEP, TmLRP, ThuLEP, and ThuVP techniques appear safe and effective with low morbidity. Comparative studies with other surgical treatments for BPH show that thulium laser prostatectomy has comparable efficacy and outcomes to TURP, PVP, and HoLEP for a wide range of prostate gland sizes and in patients taking oral anticoagulants. Also, there are lower complication and bleeding rates compared to TURP, BiTURP, and open simple prostatectomy. With more studies being conducted, thulium prostatectomy is gaining popularity and becoming part of the standard of care for treatment of BPH in prostate glands of all sizes.[7] The versatility of the thulium laser makes it a valuable tool in the armamentarium of the urologist in treating BPH. Further long-term studies are needed to support the durability of this procedure, but results thus far are highly promising.

References

1. Barbalat Y, Velez MC, Sayegh CI, Chung DE. Evidence of the efficacy and safety of the thulium laser in the treatment of men with benign prostatic obstruction. *Ther Adv Urol* 2016;**8**(3):181–91.
2. Wendt-Nordahl G, Huckele S, Honeck P, et al. Systematic evaluation of a recently introduced 2-μm continuous-wave thulium laser for vaporesection of the prostate. *J Endourol* 2008;**22**(5):1041–6.
3. Gross AJ, Netsch C, Knipper S, Hölzel J, Bach T. Complications and early postoperative outcome in 1080 patients after thulium vapoenucleation of the prostate: results at a single institution. *Eur Urol* 2013;**63**(5):859–67.
4. Wei H, Shao Y, Sun F, et al. Thulium laser resection versus plasmakinetic resection of prostates larger than 80 ml. *World J Urol* 2014;**32**(4):1077–85.
5. Macchione L, Mucciardi G, Gali' A, Di Benedetto A, Butticè S, Magno C. Efficacy and safety of prostate vaporesection using a 120-W 2-μm continuous-wave Tm:YAG laser (RevoLix 2) in patients on continuous oral anticoagulant or antiplatelet therapy. *Int Urol Nephrol* 2013;**45**(6):1545–51.
6. Pearce SM, Pariser JJ, Malik RD, et al. Outcomes following Thulium vapoenucleation of large prostates. *Int Braz J Urol* 2016;**42**(4):757–65. http://dx.doi.org/10.1590/S1677-5538.IBJU.2015.0424.
7. Gilling PJ. Laser enucleation is increasingly becoming the standard of care for treatment of benign prostatic hyperplasia of all sizes. *Eur Urol* 2013;**63**(5):868–9.
8. Fried NM, Murray KE. High-power thulium fiber laser ablation of urinary tissues at 1.94 μm. *J Endourol* 2005;**19**(1):25–31.
9. Pierce MC, Jackson SD, Dickinson MR, King TA. Laser-tissue interaction with a high-power 2-μm fiber laser: preliminary studies with soft tissue. *Lasers Surg Med* 1999;**25**(5):407–13.
10. Pierce MC, Jackson SD, Dickinson MR, King TA, Sloan P. Laser-tissue interaction with a continuous wave 3-μm fibre laser: preliminary studies with soft tissue. *Lasers Surg Med* 2000;**26**(5):491–5.

11. El-Sherif AF, King TA. Soft and hard tissue ablation with short-pulse high peak power and continuous thulium-silica fibre lasers. *Lasers Med Sci* 2003;**18**(3):139–47.
12. Mandeville J, Mourtzinos A. Surgical management of male voiding dysfunction. *Surg Clin North Am* 2016;**96**(3):491–501. http://dx.doi.org/10.1016/j.suc.2016.02.006.
13. Schomacker K, Domankevitz Y, Flotte T, Deutsch T. Co:MgF2 laser ablation of tissue: effect of wavelength on ablation threshold and thermal damage. *Lasers Surg Med* 1991;**11**(2):141–51.
14. Szlauer R, Götschl R, Razmaria A, Paras L, Schmeller NT. Endoscopic vaporesection of the prostate using the continuous-wave 2-μm thulium laser: outcome and demonstration of the surgical technique. *Eur Urol* 2009;**55**(2):368–75. http://dx.doi.org/10.1016/j.eururo.2008.10.034.
15. Hai MA, Malek RS. Photoselective vaporization of the prostate: initial experience with a new 80 W KTP laser for the treatment of benign prostatic hyperplasia. *J Endourol* 2003;**17**(2):93–6.
16. Tan AHH, Gilling PJ, Kennett KM, Frampton C, Westenberg AM, Fraundorfer MR. A randomized trial comparing holmium laser enucleation of the prostate with transurethral resection of the prostate for the treatment of bladder outlet obstruction secondary to benign prostatic hyperplasia in large glands (40 to 200 grams). *J Urol* 2003;**170**(4):1270–4.
17. Bach T, Herrmann TRW, Ganzer R, Burchardt M, Gross AJ. RevoLix™ vaporesection of the prostate: initial results of 54 patients with a 1-year follow-up. *World J Urol* 2007;**25**(3):257–62. http://dx.doi.org/10.1007/s00345-007-0171-x.
18. Xia S-J, Zhuo J, Sun X-W, Han B-M, Shao Y, Zhang Y-N. Thulium laser versus standard transurethral resection of the prostate: a randomized prospective trial. *Eur Urol* 2008;**53**(2):382–90. http://dx.doi.org/10.1016/j.eururo.2007.05.019.
19. Bach T, Wendt-Nordahl G, Michel MS, Herrmann TRW, Gross AJ. Feasibility and efficacy of Thulium:YAG laser enucleation (VapoEnucleation) of the prostate. *World J Urol* 2009;**27**(4):541–5.
20. Bach T, Herrmann TRW, Haecker A, Michel MS, Gross A. Thulium:yttrium-aluminium-garnet laser prostatectomy in men with refractory urinary retention. *BJU Int* 2009;**104**(3):361–4.
21. Netsch C, Bach T, Herrmann TRW, Gross AJ. Thulium:YAG VapoEnucleation of the prostate in large glands: a prospective comparison using 70- and 120-W 2-μm lasers. *Asian J Androl* 2012;**14**(2):325–9.
22. Netsch C, Bach T, Herrmann TRW, Neubauer O, Gross AJ. Evaluation of the learning curve for Thulium VapoEnucleation of the prostate (ThuVEP) using a mentor-based approach. *World J Urol* 2013;**31**(5):1231–8.
23. Carmignani L, Pastore AL, Picozzi SCM, et al. Thulium laser prostate enucleation in refractory urinary retention: operative and functional outcomes in a large cohort of patients. *Urology* 2016;**93**:152–7.
24. Iacono F, Prezioso D, Di Lauro G, et al. Efficacy and safety profile of a novel technique, ThuLEP (Thulium laser enucleation of the prostate) for the treatment of benign prostate hypertrophy. Our experience on 148 patients. *BMC Surg* 2012;**12**(Suppl. 1):S21.
25. Rausch S, Heider T, Bedke J, et al. Analysis of early morbidity and functional outcome of thulium: yttrium-aluminum-garnet laser enucleation for benign prostate enlargement: patient age and prostate size determine adverse surgical outcome. *Urology* 2015;**85**(1):182–8. http://dx.doi.org/10.1016/j.urology.2014.10.002.
26. Carmignani L, Bozzini G, Macchi A, Maruccia S, Picozzi S, Casellato S. Sexual outcome of patients undergoing thulium laser enucleation of the prostate for benign prostatic hyperplasia. *Asian J Androl* 2015;**17**(5):802–6.

27. Zhuo J, Bin WH, Zhao FJ, et al. Two-micrometer thulium laser resection of the prostate-tangerine technique for patients with acute urinary retention. *Lasers Med Sci* 2014;**29**(3):1093–8. http://dx.doi.org/10.1007/s10103-013-1474-3.
28. Wei HB, Zhuo J, Sun XW, et al. Safety and efficiency of thulium laser prostate resection for the treatment of benign prostatic hyperplasia in large prostates. *Lasers Med Sci* 2014;**29**(3):957–63. http://dx.doi.org/10.1007/s10103-013-1437-8.
29. Sun F, Han B, Cui D, et al. Long-term results of thulium laser resection of the prostate: a prospective study at multiple centers. *World J Urol* 2015;**33**(4):503–8.
30. Pariser JJ, Famakinwa OJ, Pearce SM, Chung DE. High-power thulium laser vaporization of the prostate: short-term outcomes of safety and effectiveness. *J Endourol* 2014;**28**(11):1357–62.
31. Vargas C, Garcia-Larrosa A, Capdevila S, et al. Vaporization of the prostate with 150-w thulium laser: complications with 6-month follow-up. *J Endourol* 2014;**28**:841.
32. Cui D, Sun F, Zhuo J, et al. A randomized trial comparing thulium laser resection to standard transurethral resection of the prostate for symptomatic benign prostatic hyperplasia: four-year follow-up results. *World J Urol* 2014;**32**(3):683–9. http://dx.doi.org/10.1007/s00345-013-1103-6.
33. Chang C-H, Lin T-P, Chang Y-H, Huang WJS, Lin ATL, Chen K-K. Vapoenucleation of the prostate using a high-power thulium laser: a one-year follow-up study. *BMC Urol* 2015;**15**:40. http://dx.doi.org/10.1186/s12894-015-0032-7.
34. Tang K, Xu Z, Xia D, et al. Early outcomes of thulium laser versus transurethral resection of the prostate for managing benign prostatic hyperplasia: a systematic review and meta-analysis of comparative studies. *J Endourol* 2014;**28**(1):65–72.
35. Zhu Y, Zhuo J, Xu D, Xia S, Herrmann TRW. Thulium laser versus standard transurethral resection of the prostate for benign prostatic obstruction: a systematic review and meta-analysis. *World J Urol* 2015;**33**(4):509–15.
36. Peng B, Wang G, Zheng J, et al. A comparative study of thulium laser resection of the prostate and bipolar transurethral plasmakinetic prostatectomy for treating benign prostatic hyperplasia. *BJU Int* 2013;**111**(4):633–7.
37. Yang Z, Wang X, Liu T. Thulium laser enucleation versus plasmakinetic resection of the prostate: a randomized prospective trial with 18-month follow-up. *Urology* 2013;**81**(2):396–401.
38. Zhang F, Shao Q, Herrmann TRW, Tian Y, Zhang Y. Thulium laser versus holmium laser transurethral enucleation of the prostate: 18-month follow-up data of a single center. *Urology* 2012;**79**(4):869–74.
39. Kyriazis I, Świniarski PP, Jutzi S, et al. Transurethral anatomical enucleation of the prostate with Tm:YAG support (ThuLEP): review of the literature on a novel surgical approach in the management of benign prostatic enlargement. *World J Urol* 2015;**33**(4):525–30.
40. Kyriazis I, Swiniarski P, Jutzi S, et al. Transurethral anatomical enucleation of the prostate with Tm:YAG support (ThuLEP): review of the literature on a novel surgical approach in the management of benign prostatic enlargement. *World J Urol* 2012;**33**:525.
41. Pereira-Correia JA, de Moraes Sousa KD, Santos JB, et al. GreenLight HPS 120-W laser vaporization vs transurethral resection of the prostate (<60 mL): a 2-year randomized double-blind prospective urodynamic investigation. *BJU Int* 2012;**110**:1184.
42. Reich O, Gratzke C, Bachmann A, et al. Morbidity, mortality and early outcome of transurethral resection of the prostate: a prospective multicenter evaluation of 10,654 patients. *J Urol* 2008;**180**:246.

Diode Laser Resection of Prostate (980 nm)

13

Brady I. Magaoay, Dominique Thomas, Bilal Chughtai
Weill Cornell Medicine, New York, NY, United States

Technical Background

Diodes are a type of semiconductor that produces monochromatic light. The material of the semiconductor, determines the final wavelength of the laser radiation. In the 1990s semiconductors were economically manufactured, leading to the proliferation and use of 1064 neodymium–yttrium–aluminum–garnet (Nd:YAG) laser systems. Early studies revealed Nd:YAG lasers to be a promising instrument to treat benign prostatic hypertrophy with comparable results to the gold-standard transurethral resection of the prostate (TURP).[1] However, the low absorption coefficient and deep optical penetration depth of the Nd:YAG caused coagulative necrosis and sloughing of tissue leading to severe storage symptoms.[2]

Contrast to Nd:YAG lasers, current diode lasers have a smaller generator size, require less energy input, and operate with electronic energy from a standard power outlet. Advanced diode lasers vary in frequency, wavelength in pulsation, power, and fiber design, which collectively affect laser–tissue interaction. Diode lasers are designed with either a front-fire or side-fire fiber design. These lasers emit light with a frequency range between 375 and 1800 nm.[4] For BPH surgery, 940, 980, 1318, and 1340 nm wavelength diode lasers are available for clinical use.[5] Power settings range from 80 to 200 W.[6] Depending on the laser wavelength and different diode models, the lasers vary in optical penetration depth from 0.5 to 7 mm.[7] *Ex vivo* models have shown tissue necrosis from 1 to 5 mm. Canine experiments have shown a necrotic zone to span upto 6 mm.[8] Ex vivo studies with bovine prostate tissue reveal that increased irradiation time and laser power increase vaporized and coagulated depths.[9]

Diode Laser Vaporization of the Prostate

Outcomes: Brief Description

Since the initial pilot studies of diode laser vaporization of the prostate (DiVAP), patients with BPH experienced significant changes in voiding parameters as well as micturition issues. A study by Seitz et al. followed 10 patients over 12 months after treatment with a 50-W 1470 nm diode laser.[10] These patients had improved International Prostate Symptom Score (IPSS), maximum urinary flow rate (Qmax), and quality of life (QoL).[10] Two cohort studies consisting of 47 and 96 men were

A Comprehensive Guide to the Prostate. https://doi.org/10.1016/B978-0-12-811464-3.00013-1

done examining of efficacy of DiVAP using a 132-W and 120-W 980 nm diode laser respectively.[11,12] Another case using a 100-W diode laser confirmed the results with significantly improved Qmax as well as postvoid residual volume (PVR) in 52 patients 12 months following the DiVAP procedure.[13] In a retrospective study of 84 men that underwent treatment with a 980-nm DiVAP for BPH.[14] The study authors found significant changes within a month after treatment with sustained improvements in the measured parameters 6 and 12 months later.[14]

In two prospective trials, researchers compared outcomes from DiVAP with a 200-W 980 nm diode to photovaporization of the prostate with a 120-W Green Light laser.[15,16] Between the two procedures, both studies showed comparable lasing time, operation time, and hospital stay.[15,16] However, patients operated with the 200-W 980 nm diode laser were exposed to greater applied energy. These two studies consisting of 139 and 117 patients had a maximum 1-year follow-up.[15,16] At the 6 and 12-month follow-up, patients experienced comparable results with significant improvements in flow parameters.[15,16] Guo et al. compared DiVAP with 980 nm 200-W diode laser to 532 nm 120-W Green Light laser with a 36-month follow-up. This study reaffirmed the short term durable outcomes.[17] A study composed of 741 patients, compared 980 nm diode laser DiVAP to 120-W Green Light HPS and thulium laser with a 2000 nm wavelength.[18] All three procedures had similar changes in Qmax, IPSS, PVR volume prostate specific antigen levels, and QoL was also compared.[18]

In one randomized controlled trial DiVAP and TURP among 72 patients for 3-months.[19] The DiVAP patients had significantly shorter catheterization times and hospital stay.[19] Patients had improved IPSS, IPSS–QoL, and Qmax, without significant difference between the two treatment groups.[19] Razzaghi et al. compared 115 patients spanning 24 months comparing 980 nm diode laser DiVAP to TURP.[20] Between the DiVAP and TURP groups, operation time was comparable, but the DiVAP patients spent significantly less time in the hospital and decreased catheter time.[20] Within the first 6 months, both groups showed similar increase in Qmax and decrease in both IPSS and PVR.[20] However, the TURP group had superior outcomes in IPSS and Qmax parameters 12 and 24 months following the procedure.[20]

In efforts to exploit the strengths of DiVAP, TURP and bipolar TURP was combined with DiVAP to enhance intraoperative hemostasis. Tzou et al. conducted a retrospective study comparing outcomes among 60 patients that were treated with either monopolar TURP or DV + bTURP.[21] For the DV + bTURP procedure, diode laser vaporization was performed first followed by bipolar TURP.[21] In accordance to other comparative studies the DV + bTURP had lower hospitalization and catheterization but greater operation time. The DV + bTURP groups had less-resected tissue as well as less post-op blood loss.[21] The 6-month follow-up presented similar IPSS, Qmax, and PVR outcomes between the TURP and DV + TURP groups.[21]

Developments in diode laser technology have produced a quartz laser tip, which enhances laser energy at the apex of the fiber without emitting a free beam. Quartz head DiVAP is a contact laser. It was originally designed to decrease the high reoperation rates following diode laser enucleation of the prostate (DiLEP) procedures due to necrotic tissue.[22] Shaker et al. conducted a randomized control trial comparing this diode laser fiber to the standard side-firing laser fiber.[23] The study consisted of 120 patients with a 6-month follow-up. The two groups had significant improvement in both voiding parameters and micturition symptoms.[23]

Safety and Complications

DiVAP has demonstrated hemostatic features as well as safe intraoperative measures. Across studies, there are no reports of capsule perforations or TURP syndrome. Collectively the commonly presented complications are with transient dysuria: 2 (20%),[11] 16 (16.7%),[12] and 11 (23.4%).[11] The 200-W 980 nm diode laser significantly reduces impaired visibility due to intraoperative bleeding.[16] Studies with lower power laser diodes (200-W) had lower rates of reoperations as well as urethral strictures.

The higher rates of reoperation following DiVAP procedures became more apparent in comparative studies. Both Chiang et al. and Guo et al. presented data with high rates of reoperation and necrotic tissue, (9.1% and 18.2%)[15] and (37.5% and 16.1%).[17] Both studies showed that the DiVAP cohort experienced high rates of reoperation and necrotic tissue development than the other treatment examined in the study.[15,17] The collected data corroborate with the ex vivo canine experiments that revealed that 980 nm 150-W diode laser caused an average necrotic zone of 6.1 mm.[22] In this study, prostatic obstructions were associated to necrotic tissue sloughing from the necrotized zones.[22] Thus the diode laser wavelength may also be factor for the high rates of obstructive necrotic tissue following DiVAP.

In contrast to these data, only four patients (8%) from the Razzaghi et al. study required reoperation, but none due to urethral stricture or bladder neck contracture.[20] The study also did not report any complications due to necrotic tissue. One caveat for this study compared to Chiang et al. and Guo et al. is that Razzaghi et al. used a 120-W diode laser for DiVAP.[15] Thus the clinical data from these studies suggest that lower powered DiVAP may reduce the rate of reoperations as well as complications due to necrotic tissue.

The quartz head fiber DiVAP, had significant decreases in dysuria and postoperative passage of tissue remnants compared to the side-firing laser fiber.[23] The quartz head fiber also showed decreased incidences of bladder neck contractures (3.6%) and urethral strictures (3.6%).[23]

Diode Laser Enucleation of the Prostate: Brief Description

Outcomes

DiLEP has several studies on efficacy and safety. Buisan et al. conducted the first DiLEP study study, this technique was similar to HoLEP.[24] 17 patients showed improvements in IPSS and Qmax with a mean hemoglobin loss of 2.1 g/dL.[24] In a second case series, which retrospectively reviewed 120 patients from a previous study, repeated the results from the first case series.[25] These patients that received DiLEP with a 980 nm diode laser expressed improved QoL and PVR. This case verified the efficacy of DiLEP treatment for patients with 60 mL prostates or larger.[25] Hruby et al. examined 43 patients undergoing DiLEP with a 1318 nm diode, which also showed improved results for patients.[26]

Other DiLEP research focused on comparing outcomes with alternative laser and surgical treatments for benign prostate hyperplasia. Lusuardi et al. conducted a prospective randomized study comparing DiLEP with a 1318 nm diode laser to bipolar TURP.[27] In this study 60 patients were divided to receive DiLEP or bipolar TURP to treat their lower urinary tract symptoms. Over the course of 6 months, the results proved that DiLEP was equally successful to bipolar TURP in improving IPSS, Qmax, and QoL. The data also revealed DiLEP to statistically significantly reduce the amount of blood loss during the procedure, catheterization time, as well as hospital stay as compared to bipolar TURP.[4] However, on average DiLEP operation time was 9 min longer than bipolar TURP to resect an equal size tissue specimen.[27] A few years later, Yang et al. completed nonrandomized comparison study examining 980 nm 80-W diode laser and TURP. DiLEP patients once again experienced less blood loss, reduced catheterization time, and decreased hospital stay in comparison to patients that underwent TURP. After 12 months, both DiLEP and TURP patients show sustained functional improvements.[28]

Researchers have also held randomized clinical trials comparing DiLEP outcomes to plasmakinetic enucleation and resection of prostate (PKERP) as well as to plasmakinetic enucleation of the prostate (PKEP). Two groups used 980 nm diode lasers for their research and reported patient outcomes (n = 80) over the course of 12 months analyzing micturition symptoms (IPSS, QoL) and voiding parameters (Qmax, PVR volume).[29,30] Both studies report improved functional outcomes over the 12 months, without any significant difference between DiLEP and PKERP.[29,30] The studies presented concurrent results in regard to perioperative outcomes. DiLEP was superior to PKERP and PKEP in regard to reduced blood loss and statistically significant reduction in catheterization time, postoperative irrigation time, hospital stay, and irritative symptoms.[29,30] These studies reported conflicting data on operation duration. The first study reported DiLEP operation time to be on average 16 min less than PKERP.[29] The more recent article published that the DiLEP procedure was statistically longer than PKEP by an average of 14.8 min.[30]

Most recently, Chen et al. studied and compared DiLEP to bipolar endoscopic enucleation of the prostate (BEEP). The studied compared perioperative results and followed 114 patients over 12 months, measuring Qmax and IPSS.[31] The researchers concluded that DiLEP was not inferior to BEEP.[31] Both DiLEP and BEEP patients experienced equal improvements in Qmax and IPSS, without any significant differences in perioperative measures.[31]

Safety and Complication

Like DiVAP, DiLEP has proven to be highly safe intraoperative procedure. From two of case studies, reoperation was necessary for a urethral stricture in 2 of 60[25] and 1 out of 17.[24] Wu et al. research included one patient from the DiLEP group that experienced a urethral stricture.[30] One case study required surgical intervention to correct a bladder neck contracture (1/129) as well as a bladder clot tamponade (1/120).[26] For Hruby et al. a needle aspiration was done on one patient for a periprostatic abscess.[26] In both cases and comparative studies, patients experienced urinary tract infections, (1/17[26], 3/120[25], and 1/30[27]). Transient incontinence was also a common complication seen first in Hruby et al.'s case study, one patient out of 17.[26] This complication was more pronounced in the comparative studies with larger patients.[27,29,30]

Among the comparative study cases, no patients required reoperations due to adenomas prostate capsule perforations. Recatheterization was observed in 5% of patients and 15% developed retrograde ejaculation.[30] 2 of 74 patients required blood transfusions in the comparative study by Yang et al. 10% of those patients incurred bladder mucosal injury, most like attributed to the learning curve for DiVAP operations.[28] A major shortcoming affecting the validity of DiVAP research is the short period of follow-up as well as the low participant volume in the studies.

Summary

In conclusion, there are various diode laser techniques that are available for prostatectomy.

Furthermore, diode laser prostatectomy is intraoperatively safe. Research studies have shown DiVAP to have a high rate or reoperation, while DiLEP has a safe adverse effects profile. Currently, there is no research evaluating the intermediate and long-term effects and complications of these procedures. Large scale randomized trials are needed to adequately evaluate its durability.

References

1. Van Melick HH, Van Venrooij GE, Eckhardt MD, Boon TA. A randomized controlled trial comparing transurethral resection of the prostate, contact laser prostatectomy and electrovaporization in men with benign prostatic hyperplasia: urodynamic effects. *J Urol* September 2002;**168**(3):1058–62.
2. Maheshwari PN, Joshi N, Maheshwari RP. Best laser for prostatectomy in the year 2013. *Indian J Urol* July 2013;**29**(3):236–43.
3. Deleted in review.
4. Lusuardi L, Mitterberger M, Hruby S, et al. Update on the use of diode laser in the management of benign prostate obstruction in 2014. *World J Urol* April 2014;**33**(4): 555–62.
5. Bach T, Muschter R, Sroka R, et al. Laser treatment of benign prostatic obstruction: basics and physical differences. *Eur Urol* February 2011;**61**(2):317–25.
6. Nair SM, Pimentel MA, Gilling PJ. A review of laser treatment for symptomatic BPH (benign prostatic hyperplasia). *Curr Urol Rep* June 2016;**17**(6):45.
7. Tholomier C, Valdivieso R, Hueber PA, Zorn KC. Photoselective laser ablation of the prostate: a review of the current 2015 tissue ablation options. *Can J Urol* October 2015;**22**(Suppl. 1):45–52.

8. Rieken M, Kang HW, Koullick E, Ruth GR, Bachmann A. Laser vaporization of the prostate in vivo: experience with the 150-W 980-nm diode laser in living canines. *Lasers Surg Med* October 2010;**42**(8):736–42.
9. Takada J, Honda N, Hazama H, Awazu K. Ex vivo efficacy evaluation of laser vaporization for treatment of benign prostatic hyperplasia using a 300-W high-power laser diode with a wavelength of 980 nm. *Laser Ther* September 30, 2014;**23**(3):165–72.
10. Seitz M, Sroka R, Gratzke C, et al. The diode laser: a novel side-firing approach for laser vaporisation of the human prostate–immediate efficacy and 1-year follow-up. *Eur Urol* December 2007;**52**(6):1717–22.
11. Erol A, Cam K, Tekin A, Memik O, Coban S, Ozer Y. High power diode laser vaporization of the prostate: preliminary results for benign prostatic hyperplasia. *J Urol* September 2009;**182**(3):1078–82.
12. Yang KS, Seong YK, Kim IG, Han BH, Kong GS. Initial experiences with a 980 nm diode laser for photoselective vaporization of the prostate for the treatment of benign prostatic hyperplasia. *Korean J Urol* November 2011;**52**(11):752–6.
13. Leonardi R. Preliminary results on selective light vaporization with the side-firing 980 nm diode laser in benign prostatic hyperplasia: an ejaculation sparing technique. *Prostate Cancer Prostatic Dis* 2009;**12**(3):277–80.
14. Kim Y, Seong YK, Kim IG, Han BH. Twelve-month follow-up results of photoselective vaporization of the prostate with a 980-nm diode laser for treatment of benign hyperplasia. *Korean J Urol* October 2013;**54**(10):677–81.
15. Chiang PH, Chen CH, Kang CH, Chuang YC. GreenLight HPS laser 120-W versus diode laser 200-W vaporization of the prostate: comparative clinical experience. *Lasers Surg Med* September 2010;**42**(7):624–9.
16. Ruszat R, Seitz M, Wyler SF, et al. Prospective single-centre comparison of 120-W diode-pumped solid-state high-intensity system laser vaporization of the prostate and 200-W high-intensive diode-laser ablation of the prostate for treating benign prostatic hyperplasia. *BJU Int* September 2009;**104**(6):820–5.
17. Guo S, Muller G, Bonkat G, et al. GreenLight laser vs diode laser vaporization of the prostate: 3-year results of a prospective nonrandomized study. *J Endourol* April 2014;**29**(4):449–54.
18. Lee WC, Lin YH, Hou CP, et al. Prostatectomy using different lasers for the treatment of benign prostate hyperplasia in aging males. *Clin Interv Aging* 2013;**8**:1483–8.
19. Cetinkaya M, Onem K, Rifaioglu MM, Yalcin V. 980-Nm diode laser vaporization versus transurethral resection of the prostate for benign prostatic hyperplasia: randomized controlled study. *Urol J* November 14, 2015;**12**(5):2355–61.
20. Razzaghi MR, Mazloomfard MM, Mokhtarpour H, Moeini A. Diode laser (980 nm) vaporization in comparison with transurethral resection of the prostate for benign prostatic hyperplasia: randomized clinical trial with 2-year follow-up. *Urology* September 2014;**84**(3):526–32.
21. Management of prostate enlargement with acute urinary retention: diode laser vaporization in combination with bipolar transurethral resection of the prostate. March 2016;**27**(1):31–34.
22. Rieken M, Bachmann A. Comparison of transurethral surgical methods. 2015.
23. Shaker HS, Shoeb MS, Yassin MM, Shaker SH. Quartz head contact laser fiber: a novel fiber for laser ablation of the prostate using the 980 nm high power diode laser. *J Urol* February 2011;**187**(2):575–9.
24. Placer J. Comment to: "Enucleation of the prostate with 980 mm diode laser (Dilep): technique and initial results". *Actas Urol Esp* January 2011;**35**(1):42–3.
25. Yang SS, Hsieh CH, Chiang IN, Lin CD, Chang SJ. Prostate volume did not affect voiding function improvements in diode laser enucleation of the prostate. *J Urol* March 2012;**189**(3):993–8.

26. Hruby S, Sieberer M, Schatz T, et al. Eraser laser enucleation of the prostate: technique and results. *Eur Urol* February 2012;**63**(2):341–6.
27. Lusuardi L, Myatt A, Sieberer M, Jeschke S, Zimmermann R, Janetschek G. Safety and efficacy of Eraser laser enucleation of the prostate: preliminary report. *J Urol* November 2011;**186**(5):1967–71.
28. Yang SS, Hsieh CH, Lee YS, Chang SJ. Diode laser (980 nm) enucleation of the prostate: a promising alternative to transurethral resection of the prostate. *Lasers Med Sci* February 2012;**28**(2):353–60.
29. Xu A, Zou Y, Li B, et al. A randomized trial comparing diode laser enucleation of the prostate with plasmakinetic enucleation and resection of the prostate for the treatment of benign prostatic hyperplasia. *J Endourol* October 2013;**27**(10):1254–60.
30. Wu G, Hong Z, Li C, Bian C, Huang S, Wu D. A comparative study of diode laser and plasmakinetic in transurethral enucleation of the prostate for treating large volume benign prostatic hyperplasia: a randomized clinical trial with 12-month follow-up. *Lasers Med Sci* May 2016;**31**(4):599–604.
31. Chen CL, Zou Z, Xu AB. Transurethral anatomical endoscopic enucleation of the prostate using diode laser versus bipolar: surgery technique with 12-month outcomes in a double-centre randomised controlled trial. *Eur Urol Suppl* March 1, 2017;**16**(3).

Open Prostatectomy

14

Michael J. Lipsky, Alexander C. Small, Carrie M. Mlynarczyk, Gina M. Badalato, Matthew P. Rutman
New York Presbyterian Hospital/Columbia University Medical Center, New York, NY, United States

Introduction

Open simple prostatectomy was long considered the gold-standard treatment for men with symptomatic benign prostatic hyperplasia (BPH) with large prostates(80 g or larger). With the development of less invasive endoscopic technology, the use of open simple prostatectomy has diminished significantly. In this chapter we review the history of open prostatectomy to understand how modern techniques were developed. The steps in the procedures are reviewed and clinical outcomes are discussed.

The Evolution of the Prostatectomy

In the late 1800s, single case reports were published describing early attempts at suprapubic (transvesical) prostatectomy with mortality rates of up to 20%–30%.[1] The first series of six patients was published in 1894 by Eugene Fuller with a mortality rate of 18%.[2] In 1912, Peter Freyer published his results with 1000 patients over his first 11 years of performing suprapubic prostatectomy.[3] In the same era, Von Stockum described the retropubic prostatectomy approach to avoid opening the bladder and improve visualization of the distal adenoma and sphincter. Instead, he incised the prostatic capsule to core out the adenomatous tissue, packed the cavity, and drained the bladder with a suprapubic tube.[4]

Over the next several decades, technical modifications were made to the procedure to decrease morbidity and mortality rates. For example, in 1945 Millin improved the retropubic approach by creating a watertight closure of the prostatic capsule to allow urethral catheter drainage.[5] In 1965, Malament noted postoperative hemostatic improvement using a removable purse-string suture separating the bladder from the prostatic fossa.[6] This technique was subsequently shown to have favorable results in a large series and is still used in contemporary practice.[7,8]

Compared to transurethral approaches, simple prostatectomy has been shown to be a more definitive treatment in several domains, most notably a lower retreatment rate (1.8%–4.5% vs. 12%–15.5%).[9] However, with advances in medical therapies and less invasive approaches, open simple prostatectomy has become less commonly utilized. By 1985, 95% of prostatectomies in Medicare patients were performed via a transurethral approach.[9] Indeed, a recent study of the National Inpatient Sample database from 2002 to 2012 showed an average annual decrease in simple prostatectomy cases by 145 cases per year with a concurrent increase in the proportion of minimally invasive

A Comprehensive Guide to the Prostate. https://doi.org/10.1016/B978-0-12-811464-3.00014-3

approaches from 0% in 2002 to 5% in 2012.[10] Reasons for this decline include significant intraoperative blood loss, long hospitalization, and improvements in alternative treatments. Despite this trend, contemporary studies continue to demonstrate significant improvement in urinary symptoms and quality of life after open prostatectomy.[11,12] In selected men, open prostatectomy can be a well-tolerated and highly effective operation. Additionally, simple prostatectomy continues to be an important treatment in the developing world where minimally invasive treatments and medical therapies are not as readily available.[8]

Modern evolution of minimally invasive surgery has also dramatically changed the role of open prostatectomy. In 1997, Schuessler reported the first series of laparoscopic radical prostatectomy for prostate cancer.[13] As urologists became more experienced with minimally invasive surgery, techniques for laparoscopic and robotic simple prostatectomy were developed. In 2002, the first laparoscopic simple prostatectomy was performed on a 173-g prostate using a modified retropubic approach.[14] As this minimally invasive approach became more popular, one small study was performed to compare laparoscopic to open retropubic prostatectomy, and showed only decreased blood loss in the laparoscopic procedure.[15] In 2008, the robotic approach was introduced for simple prostatectomy.[16] Since that time, further adjustments have been made to decrease perioperative complications and maximize adenomatous tissue removed with these minimally invasive approaches.

Indications for Open Prostatectomy

Indications for surgery for BPH are described in the American Urological Association (AUA) and EAU guidelines.[17,18] Surgery is recommended for patients who have renal insufficiency secondary to BPH, those who have recurrent urinary tract infections, bladder stones, or severe gross hematuria due to BPH and those who have lower urinary tract symptoms (LUTS) refractory to other therapies. There are a number of reasons to consider simple prostatectomy over other treatment options. These include men with large prostates (generally described as greater than 80 g), plan for concurrent bladder procedures such as diverticulectomy or cystolithotomy, and in patients who are unable to be placed in dorsal lithotomy position due to severe hip disease.[18]

Procedure

Perioperative Evaluation and Management

Before prostatectomy, a full urologic workup begins with evaluation of LUTS. Alternative causes of LUTS including urethral stricture, bladder stones, and/or bladder diverticula should be ruled out before proceeding to surgery. A thorough history and physical exam combined with office cystoscopy can identify most of these causes. Likewise, if patients have history of gross or microscopic hematuria, a full workup with cystoscopy and upper tract evaluation (CT urogram or renal bladder ultrasound plus retrograde pyelograms) should be performed. Cystoscopy should also note urethral length and presence of a median lobe. Finally, it is essential to screen for prostate cancer as per appropriate regional guidelines with prostate specific antigen, digital rectal exam and, if indicated, transrectal ultrasound-guided biopsy. Ultrasound is not required in of itself before prostatectomy, but can be useful to determine prostate size and its impact on type of surgical approach.

Since patients with BPH are often elderly, many are affected by comorbid conditions such as hypertension, chronic obstructive pulmonary disease, coronary artery disease, etc. Careful preoperative assessment in conjunction with an internist is necessary, and further cardiac evaluation may be considered for patients at elevated risk. Also, a detailed medication review is required with a focus on drugs that affect bleeding such as aspirin, clopidogrel (Plavix), warfarin (Coumadin) or next-generation anticoagulants such as dabigatran (Pradaxa), rivaroxaban (Xarelto), or apixaban (Eliquis). These medications must be discontinued at the time of surgery, and perioperative bridging may require assistance from a cardiologist or hematologist.

For men with chronic urinary retention, renal evaluation may be needed with serum creatinine measurement. If elevated, renal bladder ultrasound can reveal urinary bladder wall thickening and hydronephrosis that may prompt catheter drainage until serum creatinine nadirs. Those in retention also experience urinary stasis, which puts them at increased risk for recurrent urinary tract infections (UTI) and bladder stones.

Patients should be fully counseled on surgical risks before surgery. One notable complication after open prostatectomy is bleeding requiring blood transfusion, which occurs in approximately 0%–36% of men. Therefore, patients require a recent blood sample for blood typing and crossmatching by the hospital's blood bank. Some patients may elect to bank their own blood in the weeks leading up to surgery for autologous transfusion if required. Another complication that can occur is rectal injury during prostate enucleation. Surgeons can consider administering a bowel preparation the day before surgery. Deep vein thrombosis (DVT) is a risk with any prolonged immobilization during and after surgery, so pharmacologic prophylaxis with heparin or enoxaparin (Lovenox) should be administered along with lower extremity compression stockings and/or sequential compression devices.

Retropubic Prostatectomy

The patient should be placed in the supine position on the operating table. The table is flexed to increase exposure to bladder and prostate by maximizing distance between the pubis and umbilicus. The patient is prepped and draped in standard fashion and perioperative antibiotics are administered. If not previously performed, rigid or flexible cystoscopy can be done at this point. Next, the table is then tilted 15–30°degrees into the Trendelenburg position to shift intraabdominal viscera cephalad.

Incision is made in the lower midline and extended 10 cm from below the umbilicus to 2–3 cm above the pubic symphysis. Alternatively, a Pfannenstiel incision can be utilized. The incision is extended through the anterior rectus sheath and the rectus abdominus muscles are separated laterally in the midline. Staying extraperitoneal, the Space of Retzius is entered and the peritoneum is gently swept cephalad. A self-retaining retractor (i.e., Balfour, Omni Tract, etc.) is placed with a malleable blade in the superior position to retract the bladder and expose the anterior prostate.

The catheter balloon should not be retracted to ease identification of the bladder neck. With the prostate visualized, the periprostatic fat is gently dissected away to expose the dorsal venous complex (DVC) and puboprostatic ligaments.

At this point, hemostatic maneuvers should be taken to limit bleeding once prostate enucleation begins. The blood supply to the prostate comes from the prostatovesical arteries that divide at the base of the bladder to form the inferior vesical and prostatic arteries at the lateral pedicles of the prostate. With a BPH adenoma, the penetrating (or urethral) branches from the posterolateral prostatic artery can grow in size and briskly bleed if not properly controlled. First, the DVC is ligated (3-0 monocryl on 5/8 inch circle tapered "GU" needle). Second, the endopelvic fascia is incised bilaterally lateral to the puboprostatic ligaments and the lateral pedicles are exposed. The pedicles are ligated with deep figure of eight sutures (2–0 chromic on CTX needle) placed posterolaterally to the prostate at the junction of the seminal vesicles. Care must be taken to avoid the neurovascular bundles in this area.

The prostate enucleation can be performed through a transverse capsular incision (Millin Technique) or alternatively through a vesicocapsular technique. To begin the transverse capsular approach, two parallel transverse rows of stay sutures (2-0 chromic) are placed deeply through the anterior prostate capsule into the adenoma 1.5–2 cm distal to the bladder neck near the apex of the prostate. The foley catheter is removed, and the capsulotomy is made between the parallel suture lines using electrocautery or a sharp blade. The upper portion of the capsule can be compressed with a sponge stick at this point to allow fulguration of bleeding vessels. Enucleation is started by identifying a plane between the capsule and adenoma by gently using an index finger to dissect bluntly and circumferentially. Care should be taken along the posterior capsule since rectal injury can occur. If a tear occurs between the prostate and rectum, it can often be closed in two layers with an interposing layer of omentum, but for a large tear or one with stool spillage, a colorectal surgeon is typically consulted for a diverting colostomy. When the adenoma has been dissected, the anterior commissure of the adenoma should be divided with scissors from the bladder neck to apex to expose the prostatic urethra and allow for division at the level of the verumontanum proximal to the external sphincter. The adenoma can be removed with a Babcock clamp and any residual tissue or median lobe should be removed.

The alternative approach to enucleation is the vesicocapsular technique, where a vertical incision is made through the midline of the bladder neck and prostate to approach the adenoma. Enucleation is performed in a similar circumferential blunt fashion and the adenoma is divided near the apex of the capsular incision. The approach incorporates the anterior bladder wall and prostatic capsule as one continuous incision. It creates an increased risk for inadvertent extension of the incision into the external urinary sphincter and subsequent urinary leakage after surgery. The transverse capsular (Millin) technique is consequently often favored.

Following enucleation, hemostasis and bladder neck reconstruction are performed. Traditionally after the adenoma is removed, the prostatic fossa was packed with moist gauze for 5 min to compress bleeding vessels and then the capsule was closed. Now combined pressure and selective electrocautery and/or suture ligation is favored for maximal hemostasis. Two figure of eight sutures (4-0 chromic) are placed at the 5- and

7-o'clock positions to control any large penetrating branches from the posterolateral prostatic arteries. Next the urethral orifices are identified to ensure safe distance from the bladder neck before closure. The "trigonalization maneuver" is then performed by bringing the posterior lip of the bladder neck distally into the prostatic fossa and suturing it in place (2-0 chromic). This maneuver aids hemostasis, speeds epithelization of the fossa, and reduces risk of bladder neck contracture.

Finally, a large bore three-way foley catheter (22- or 24-french) is placed per urethra into the bladder. The anterior capsule is closed with two running sutures (2-0 chromic on 5/8 in. circle tapered "GU" needle) from each lateral side of the capsulotomy and joined in the midline to create a watertight seal. The foley balloon is inflated with 30–50 mL sterile water and pulled back to the bladder neck to tamponade the fossa. The bladder is irrigated and closure checked for leakage. A closed system drain (i.e., Jackson Pratt, Davol, etc.) is left in the Space of Retzius anterior to the capsulotomy and is brought to the skin through a separate site. The rectus fascia and skin are closed in the standard fashion and the wound is dressed. Before transfer of the patient to the recovery room, the foley catheter is secured and connected to continuous bladder irrigation.

Suprapubic (Transvesical) Prostatectomy

The suprapubic (transvesical) approach allows for an excellent view of the bladder neck and ureteral orifices before enucleation and therefore can be better for men with a large median lobe, bladder stones, or bladder diverticula. Patient positioning, approach, and hemostatic maneuvers are performed in the same manner as the retropubic approach. Instead of entering the prostate through the capsule, the bladder is entered through a 4–5 cm anterior midline vertical cystotomy. Stay sutures (3-0 vicryl) are placed on either side of the incision, which is made with electrocautery. A figure of eight stitch is often placed at the inferior aspect of the incision to prevent extension into the prostate capsule. A narrow malleable retractor with moist gauze is placed into the bladder to retract it cephalad and connected to the self-retaining retractor. After identifying the ureteral orifices, electrocautery is used to cut the bladder epithelium around the bladder neck at the protruding prostate. Curved scissors are used to develop a plane between the adenoma and capsule, and then an index figure is swept circumferentially to bluntly dissect the adenoma. At the apex, the prostatic urethra is transected bluntly with a pinching motion between the thumb and index figure. The adenoma is removed and the fossa is inspected for residual tissue. Similar to the retropubic approach, hemostasis is achieved in the fossa and the trigonalization maneuver is performed. A large bore three-way foley catheter (22- or 24-french) is placed, or alternatively a urethral catheter and Malecot suprapubic tube can be used. The cystotomy is closed in two layers with a running suture (2-0 vicryl) followed by joining the previously placed stay sutures to form a watertight seal. If bleeding persists, a temporary purse string suture (1-0 nylon or polydioxanone synthetic absorbable suture) can be placed around the posterior margin of the bladder neck through mucosa into the muscle and tightened around the catheter. Both ends of that suture are brought through the body wall and tied together over a gauze pad to allow for removal 1 or 2 days after surgery. A drain is placed, the surgical site is closed, and the catheter is connected to irrigation.

Postoperative Management

Postoperatively, patients are maintained on continuous bladder irrigation and IV hydration until their urine clears, typically in 2–3 days. If bleeding continues, the catheter can be placed on traction to tamponade the prostatic fossa and blood transfusion may be required. If bleeding remains uncontrolled, transurethral fulguration can be attempted. The drain is left in place until the output slows, and finally foley catheter is removed usually 3–5 days after drain removal. DVT prophylaxis is continued while the patient is hospitalized and antibiotics are continued until catheter is removed for trial of void.

Outcomes

Open Prostatectomy Success Rates

The success of interventions for BPH may be defined in a number of different ways including changes in patient-driven symptom scores such as the International Prostate Symptom Score (IPSS), objective changes in metrics of obstruction such as peak urinary flow rates (Qmax) or postvoid residual urine (PVR), and/or by the need for repeat interventions. Historical success rates of open simple prostatectomy have been variable as the definitions of success differ between studies and depend on several factors such as timing of postoperative assessment measures. The body of evidence for open prostatectomy consists primarily of single institutional case series (Table 14.1), as well as series comparisons of prostatectomy to newer endoscopic approaches. In addition to the procedural efficacy, other outcome measures are important when comparing minimally invasive and open surgical procedures. These include operative time, hospital length of stay, postoperative pain, and postoperative complications.

Changes in Patient Symptom Scores

While historically, much of the data on open prostatectomy focused on the morbidity of the operation, more recent studies add data on improvements in validated patient symptom scores. Since the introduction of the AUA symptom index and the IPSS, these scores have been used to quantify patient BPH symptoms. Given the ability to objectively track patient symptoms, these scores have been used as a marker of success of procedures for BPH surgery. Of note, a three-point improvement is needed to be considered a meaningful change.[18] As one of the indications for prostatectomy for BPH is a symptom refractory to medical intervention, the change in patient symptoms is of utmost importance. In a small prospective case series of 32 patients, Tubaro et al. noted a profound improvement in IPSS score of about 18.5 points (93%) from a baseline of 19.9 points. They also found a 4.7 point improvement (96%) in the quality of life score at 1 year postoperatively.[19] Similarly, Gacci et al. described reduction of 12 points (62%) in IPSS after prostatectomy using a transvesical approach with 6 months of follow-up.[11] In a retrospective review of a series of patients with large prostates (mean prostate size 104.5 mL), significant improvements in prostate-related

Table 14.1 **Efficacy rates of open prostatectomy by study**

Study	Approach	Patients (N)	Mean prostate volume (mL)	Hospital stay/ catheterization (days)	ΔIPSS (% change)	ΔQmax, mL/s (% change)	ΔPVR, mL (% change)	Reoperation rate (%)
Tubaro et al.[19]	Transvesical	32	84	6.2/5.4	−18.5 (−93%)	+19.9 (219%)	−120 (−94%)	–
Kuntz et al.[23]	Transvesical	60	113	10.5/8.1	−18.2 (−87%)	+23.4 (+650%)	−290 (99%)	–
Serretta et al.[27]	88.8% Transvesical	1804	75	7/5	–	–	–	3.6%
Gacci et al.[11]	Transvesical	60	–	6/5.5	−12 (62%)	+17.9 (+218%)	−92 (−91%)	–
Adam et al.[24]	Transvesical	201	90.3	10/6	–	+14.8 (+247%)	−85 (−100%)	–
Shaheen et al.[32]	92% Retropubic	37	97.8	11.7	–	–	–	24.3%
Varkarakis et al.[12]	Transvesical	232	104.5	6/5	−23.3 (−93%)	+16.5 (+226%)	−105(-88%)	3.9%
Helfand et al.[20]	75% Retropubic	56	138.5	6.4/4.7	−14.5 (−78%)	–	−297 (−94%)	–
Gratzke et al.[25]	–	902	96.3	11.9	–	+12.5 (+118%)	−128 (−88%)	–
Geavlete et al.[21]	Transvesical	70	130	6.9/5.8	−21.3 (−83%)	+19.4 (+340%)	−147 (−88%)	–
Rao et al.[22]	Transvesical	40	110	9.3/6.2	−20.4 (−83%)	+19.5 (+331%)	−76 (93%)	5.3%

ΔIPSS, change in IPSS score; *ΔPVR*, change in postvoid residual urine; *ΔQmax*, change in peak flow rate.

symptoms were found with a decrease in an average of 23.3 points (93%) on the IPSS survey with quality of life improvements of 4.1 points (87%).[12] Overall the improvements in IPSS scores range from 62% to 93% with a mean improvement of 83% (Table 14.1).[11,12,19–23]

Changes in Obstructive Parameters

While symptom assessment is one way to determine the efficacy of intervention for BPH, objective measures may be used to gage improvement in urine flow. That is, increase in maximum flow rate on uroflowmetry and decrease in postvoid residual urine in the bladder. The change in Qmax ranges from about +12.5 mL/s to about +20 mL/s.[11,12,19,21–25] In a large multicenter prospective case series from Germany including 902 patients, the mean increase in maximum flow rate was about 12.5 mL/s from a baseline of 10.6 mL/s preoperatively to 23.1 mL/s postoperatively.[25] At the same time, the mean postvoid residual urine volume decreased from a mean 145 to 17.5 mL for an average decrease of 127.5 mL. The range of decrease in PVR varies from 85 to 297 mL corresponding to a mean change of 93% from the baseline PVR.

Reoperation

Measurement of reoperation rate after surgery for BPH is time-dependent and therefore lengthy follow-up may be necessary. Reoperation includes repeat interventions for symptoms of obstruction unrelieved by the initial surgery, surgery for obstructive sequelae of the initial procedure (i.e., stricture disease, bladder neck stenosis), or prostatic regrowth leading to further obstruction. For the definition of success, this would not include surgery performed due to uncontrolled bleeding. In 1989, Roos et al. published a population-based series comparing transurethral resection of the prostate (TURP) to open prostatectomy including about 15,000 prostatectomies from Europe and Canada. With 8 years of follow-up, 3.9% (regional range 1.8%–4.5%) of patients underwent a second prostatectomy (either open or transurethral).[9] In another population-based cohort, Semmens et al. analyzed the records of Australian men who underwent either TURP or open prostatectomy from 1980 to 1995, including 1134 open procedures. At 8 years of follow-up, the cumulative incidence of first repeat prostatectomy was 3.3% in patients who underwent open prostatectomy as compared to the 6.6% of patients who initially underwent TURP.[26] Similarly, Serretta et al. analyzed more than 1800 prostatectomies in Italy over a 2-year period and found the reoperation rate to be 3.6% within 2 years of surgery.[27] The reasons for these surgeries were mainly due to bladder neck and/or urethral stenosis. Finally, an analysis of an Austrian population cohort from 1992 to 1996 including 2452 open prostatectomies demonstrated a 9.5% cumulative incidence of repeat endourological intervention (including TURP, urethrotomy, bladder neck incision) within 8 years of primary surgery. When limiting this number to need for a secondary TURP only, this figure was 3.4% at years.[28] Overall, it appears that open prostatectomy is an effective treatment approach with a need for further intervention for BPH symptoms <5%.

Morbidity

Complication rates vary widely throughout the literature on open prostatectomy. In an early series of transvesical prostatectomy described by Fuller in 1895, mortality rates were found to occur in up to 18% of cases.[29] This decreased to 5.5% in a review of Freyer's first 1000 cases.[3] With improvements in techniques as well as anesthesia improvements, complication rates in contemporary series have declined substantially, with a negligible mortality rate (Table 14.2). Complications of open prostatectomy may be divided into immediate and delayed complications. The most commonly cited immediate postoperative complication is bleeding, often leading to transfusion. This is found in 0%–36% of cases in the current literature.[8,19,20,24,25,27,30–33] Perioperative mortality has been cited in up to 3% of cases[32] but is generally <1%.[9,28] In a review of the National Inpatient Sample including >35,000 prostatectomies, the rate of perioperative mortality was found to be 0.4%.[10] While a number of late complications have been described, the three most commonly reported complications include bladder neck stenosis, urinary incontinence, and urethral stricture. On review of >1800 patients the rate of urethral and bladder neck stenosis was found to be 4.8% while the rate of urinary incontinence was 1.2% within 2 years of surgery.[27] Overall, the incidence of late complications is low.

Comparing Open Prostatectomy Approaches

While numerous studies exist with each approach to open prostatectomy, there has only been one randomized trial comparing the two main techniques for open prostatectomy. 65 patients with prostate volume >75 g were randomized to retropubic prostatectomy (n = 32) or suprapubic prostatectomy (n = 33). While both approaches were effective improving urinary flow rate and IPSS scores, there was a higher rate of complications with the suprapubic approach and the retropubic approach had a significantly higher maximum flow rate postoperatively. However, when limiting this subset to prostates >100 g, the retropubic approach had a higher blood loss.[34]

Comparing Open Prostatectomy to Other Surgical Treatments for Benign Prostatic Hyperplasia

Open prostatectomy has been shown to be an effective treatment for symptomatic BPH and its safety profile has improved over the last century. However, as new therapies were introduced or modified, their efficacy and complications were compared to that of open prostatectomy. As TURP has been around since before 1900 and has evolved significantly since that time to the contemporary procedure performed,[4] there are robust data comparing TURP and open prostatectomy. Meyhoff and Nordling conducted a randomized trial of 75 patients comparing transurethral and transvesical prostatectomy. They found that both groups experienced significant symptomatic improvement as well as improvements in urinary flow. The overall patient dissatisfaction rate was 16% in the TURP group and 9% in the open prostatectomy group.

Table 14.2 Complications of open prostatectomy

Study	Patients (N)	Early complications			Late complications		
		Overall (%)	Transfusion (%)	Death (%)	Bladder neck stenosis (%)	Urethral stricture (%)	Incontinence (%)
Roos et al.[9]	14,990	–	–	2.5	–	–	–
Meier et al.[8]	240	20	4.6	0	1.7	–	–
Condie et al.[30]	200	8	1	1	1	–	0.5
Tubaro et al.[19]	32	31	0	0	6.25	–	9.4
Kuntz et al.[23]	60	23.4	13.3	1.7	3.3	–	–
Hill et al.[31]	106	32	4.7	0.9	–	–	–
Serretta et al.[27]	1804	37	8.2	0.06	4.8	–	1.2
Adam et al.[24]	201	23	18.9	0	–	–	0.5
Shaheen et al.[32]	37	43	16.2	3	5.4	8.1	8.1
Varkarakis et al.[12]	232	17	6.8	0	3.3	0.6	0
Helfand et al.[20]	56	41	36	0	5.4	5.4	5.4
Zargooshi et al.[33]	3000	6	3.3	0.1	–	–	–
Gratzke et al.[25]	902	17.3	7.5	0.2	–	–	–
Elshal et al.[39]	163	42.3	24.5	1.2	–	–	–
Geavlete et al.[21]	70	21	8.6	0	1.6	3.2	1.6
Rao et al.[22]	40	40	10	0	5.3	7.9	
Pariser et al.[10]	35,171	28	24	0.4	–	–	–

About 25% of patients in each group had secondary operations for BPH at 5-year follow-up.[35] However, data from large population-based studies suggest that TURP has a higher rate of reoperation than open surgery.[9,26,28] While some studies have and increased mortality rate associated with TURP,[9] others have found no difference between the two operations.[28] TURP is generally considered first line therapy for prostates <80 g but is only an alternate therapy for larger glands.

For prostate glands larger than 80 g, the primary treatment options include either open prostatectomy or holmium laser enucleation of the prostate (HoLEP).[17] Rather than resecting prostate tissue, HoLEP enucleates adenomatous tissue, making it more similar to open surgery than TURP. A metaanalysis by Jones et al. comparing HoLEP to open prostatectomy found no difference in Qmax, PVR, IPSS, or quality of life scores at 12 or 24 months. Open prostatectomy was associated with shorter operative time and greater volume of tissue removed, but also with increased blood loss, longer catheterization time, and longer hospital stay.[36] The data regarding endoscopic enucleation are further strengthened by a metaanalysis including all forms of endoscopic enucleation (HoLEP, bipolar plasma vaporization enucleation of the prostate, and plasmakinetic enucleation of the prostate).[37] Again, no difference was noted between the two groups in IPSS, quality of life score, Qmax, or PVR. Endoscopic enucleation was noted to once again have decreased catheterization time and hospital stay. Open prostatectomy was associated with more blood loss and transfusions, but shorter operating time. Given its similar functional outcomes, with decreased hospitalization, endoscopic enucleation is often considered a desirable alternative to open surgery.

With advances in minimally invasive surgery, simple prostatectomy has been performed using laparoscopic and robotic approaches. Given their novelty, however, there were insufficient data to make any formal recommendation in the recently updated AUA guidelines, and their use is denoted as an option.[18] Porpiglia et al. conducted a prospective nonrandomized study of 40 patients comparing laparoscopic to open prostatectomy.[15] This study found a significantly longer operative time with the laparoscopic approach but with decreased blood loss. No difference was noted in overall complication rate, hospital stay or postoperative parameters at 2 month follow-up. As many of the studies utilizing minimally invasive techniques are small case series, a pooled analysis of 487 robotic and 843 laparoscopic simple prostatectomies was conducted in 2015.[38] With a median prostate volume of 100 mL, median blood loss was 200 mL, hospital stay was 4 days, and the complication rate was 10.6%. Lucca et al. conducted a metaanalysis comparing minimally invasive and open prostatectomy. Minimally invasive surgery was associated with shorter hospital stay, shorter catheterization time, and less blood loss than open surgery. Open prostatectomy was found to have shorter operative times. Functional outcomes including IPSS and Qmax at follow-up were similar between the two groups. Given the similar results between the open and minimally invasive approaches, combined with the decreased hospitalization time, these therapies warrant further inspection. As endoscopic enucleation, open prostatectomy, and minimally invasive prostatectomy appear to be effective, further studies are likely needed to determine ideal treatment for symptomatic men with large prostates.

Conclusions

Simple prostatectomy is an effective treatment for BPH and had historically been the gold standard for large prostate glands. Improvements in medical management and the advancement of minimally invasive techniques have led to decreased use of open simple prostatectomy. However, it continues to be a safe and effective option for the treatment symptomatic BPH, especially in patients with large prostate glands.

References

1. Ricketts BM. Surgery of the prostate, pancreas and diaphragm. *Cincinnati Lancet-Clin* 1902;**88**:399–405.
2. Fuller E. Six successful and successive cases of prostatectomy. *J Cutan Genito-Urin Dis* 1895;**13**:229.
3. Freyer P. One thousand cases of total enucleation of the prostate for radical cure of enlargement of that organ. *Br Med J* 1912:868–70.
4. Herr HW. The enlarged prostate: a brief history of its surgical treatment. *BJU Int* 2006;**98**(5):947–52.
5. Millin T. Retropubic prostatectomy; a new extravesical technique; report of 20 cases. *Lancet* 1945;**2**(6380):693–6.
6. Malament M. Maximal hemostasis in suprapubic prostatectomy. *Surg Gynecol Obstet* 1965;**120**:1307–12.
7. Nicoll GA, Riffle 2nd GN, Andersen FO. Suprapubic prostatectomy. The removable purse string: a continuing comparative analysis of 300 consecutive cases. *J Urol* 1978;**120**(6):702–4.
8. Meier DE, Tarpley JL, Imediegwu O, et al. The outcome of suprapubic prostatectomy: a contemporary series in the developing world. *Urology* 1995;**46**:40–4.
9. Roos NP, Wennberg JE, Malenka DJ, et al. Mortality and reoperation after open and transurethral resection of the prostate for benign prostatic hyperplasia. *NEJM* 1989;**320**(17): 1120–4.
10. Pariser JJ, Pearce SM, Patel SG, Bales GT. National trends of simple prostatectomy for benign prostatic hyperplasia with an analysis of risk factors for adverse perioperative outcomes. *Urology* 2015;**86**(4):721–5.
11. Gacci M, Bartoletti R, Figlioli S, et al. Urinary symptoms, quality of life and sexual function in patients with benign prostatic hypertrophy before and after prostatectomy: a prospective study. *BJU Int* 2003;**91**:196–200.
12. Varkarakis I, Kyriakakis Z, Delis A, Protogerou V, Deliveliotis C. Long-term results of open transvesical prostatectomy from a contemporary series of patients. *Urology* 2004;**64**(2):306–10.
13. Schuessler WW, Schulam PG, Clayman RV, Kavoussi LR. Laparoscopic radical prostatectomy: initial short-term experience. *Urology* 1997;**50**(6):854–7.
14. Mariano MB, Graziottin TM, Tefilli MV. Laparoscopic prostatectomy with vascular control for benign prostatic hyperplasia. *J Urol* 2002;**167**(6):2528–9.
15. Porpiglia F, Terrone C, Renard J, et al. Transcapsular adenomectomy(Millin): a comparative study, extraperitoneal laparoscopy versus open surgery. *Eur Urol* 2006;**49**(1):120–6.
16. Sotelo R, Clavijo R, Carmona O, et al. Robotic simple prostatectomy. *J Urol* 2008; **179**(2):513–5.

17. Oelke M, Bachmann A, Descazeaud A, et al. EAU guidelines on the treatment and follow-up of non-neurogenic male lower urinary tract symptoms including benign prostatic obstruction. *Eur Urol* 2013;**64**(1):118–40.
18. McVary KT, Roehrborn CG, Avins AL, et al. Update on AUA guideline on the management of benign prostatic hyperplasia. *J Urol* 2011;**185**(5):1793–803.
19. Tubaro A, Carter S, Hind A, Vicentini C, Miano L. A prospective study of the safety and efficacy of suprapubic transvesical prostatectomy in patients with benign prostatic hyperplasia. *J Urol* 2001;**166**:172–6.
20. Helfand B, Mouli S, Dedhia R, McVary KT. Management of lower urinary tract symptoms secondary to benign prostatic hyperplasia with open prostatectomy: results of a contemporary series. *J Urol* 2006;**176**(6 Pt 1):2557–61. discussion 2561.
21. Geavlete B, Stanescu F, Iacoboaie C, Geavlete P. Bipolar plasma enucleation of the prostate vs open prostatectomy in large benign prostatic hyperplasia cases – a medium term, prospective, randomized comparison. *BJU Int* 2013;**111**(5):793–803.
22. Rao JM, Yang JR, Ren YX, He J, Ding P, Yang JH. Plasmakinetic enucleation of the prostate versus transvesical open prostatectomy for benign prostatic hyperplasia >80 mL: 12-month follow-up results of a randomized clinical trial. *Urology* 2013;**82**(1): 176–81.
23. Kuntz RM, Lehrich K. Transurethral holmium laser enucleation versus transvesical open enucleation for prostate adenoma greater than 100 gm.: a randomized prospective trial of 120 patients. *J Urol* 2002;**168**(4 Pt 1):1465–9.
24. Adam C, Hofstetter A, Deubner J, et al. Retropubic transvesical prostatectomy for significant prostatic enlargement must remain a standard part of urology training. *Scand J Urol Nephrol* 2004;**38**:472–6.
25. Gratzke C, Schlenker B, Seitz M, et al. Complications and early postoperative outcome after open prostatectomy in patients with benign prostatic enlargement: results of a prospective multicenter study. *J Urol* 2007;**177**(4):1419–22.
26. Semmens J, Wisniewski Z, Bass A, Holman C, Rouse I. Trends in repeat prostatectomy after surgery for benign prostate disease: application of record linkage to healthcare outcomes. *BJU Int* 1999;**84**:972–5.
27. Serretta V, Morgia G, Fondacaro L, et al. Open prostatectomy for benign prostatic enlargement in southern Europe in the late 1990s: a contemporary series of 1800 interventions. *Urology* 2002;**60**:623–7.
28. Madersbacher S, Lackner J, Brossner C, et al. Reoperation, myocardial infarction and mortality after transurethral and open prostatectomy: a nation-wide, long-term analysis of 23,123 cases. *Eur Urol* 2005;**47**(4):499–504.
29. Tubaro A, de Nunzio C. The current role of open surgery in BPH. *EAU-EBU Update Ser* 2006;**4**(5):191–201.
30. Condie JD, Cutherell L, Mian A. Suprapubic prostatectomy for benign prostatic hyperplasia in rural Asia: 200 consecutive cases. *Urology* 1999;**54**(6):1012–6.
31. Hill A, Njoroge P. Suprapubic transvesical prostatectomy in a rural Kenyan hospital. *East Afr Med J* 2002;**79**(2):65–7.
32. Shaheen A, Quinlan D. Feasibility of open simple prostatectomy with early vascular control. *BJU Int* 2004;**93**:349–52.
33. Zargooshi J. Open prostatectomy for benign prostate hyperplasia: short-term outcome in 3000 consecutive patients. *Prostate Cancer Prostatic Dis* 2007;**10**(4):374–7.
34. Carneiro A, Sakuramoto P, Wroclawski ML, et al. Open suprapubic versus retropubic prostatectomy in the treatment of benign prostatic hyperplasia during resident's learning curve: a randomized controlled trial. *Int Braz J Urol* 2016;**42**(2):284–92.

35. Meyhoff HH, Nordling J. Long term results of transurethral and transvesical prostatectomy. A randomized study. *Scand J Urol Nephrol* 1986;**20**(1):27–33.
36. Jones P, Alzweri L, Rai BP, Somani BK, Bates C, Aboumarzouk OM. Holmium laser enucleation versus simple prostatectomy for treating large prostates: results of a systematic review and meta-analysis. *Arab J Urol* 2016;**14**(1):50–8.
37. Li M, Qiu J, Hou Q, et al. Endoscopic enucleation versus open prostatectomy for treating large benign prostatic hyperplasia: a meta-analysis of randomized controlled trials. *PLoS One* 2015;**10**(3):e0121265.
38. Autorino R, Zargar H, Mariano MB, et al. Perioperative outcomes of robotic and laparoscopic simple prostatectomy: a European-American multi-institutional analysis. *Eur Urol* 2015;**68**(1):86–94.
39. Elshal AM, El-Nahas AR, Barakat TS, Elsaadany MM, El-Hefnawy AS. Transvesical open prostatectomy for benign prostatic hyperplasia in the era of minimally invasive surgery: perioperative outcomes of a contemporary series. *Arab J Urol* 2013;**11**(4):362–8.

Robotic-Assisted Laparoscopic Surgery

15

Sven J. Walderich, Dominique Thomas, Bilal Chughtai
Weill Cornell Medicine, New York, NY, United States

Introduction

Benign prostatic hyperplasia (BPH) is characterized by prostatic stromal and epithelial cell hyperplasia resulting in the formation of discrete nodules in the periurethral region of the prostate.[1] BPH is the most common benign neoplasm among men affecting nearly 6.5 million men between the ages of 50 and 79 in the United States.[2,3] Histologic evidence of BPH can be identified in approximately 90% of men over the age of 80 years.[4–6] While most many patients with BPH are asymptomatic, 40% of men will experience clinically significant lower urinary tract symptoms (LUTS), which are characterized as either obstructive or irritative symptoms. Obstructive symptoms of BPH include urinary hesitancy, weak flow, prolonged voiding, and urinary retention, which can result in overflow incontinence. In contrast, irritative symptoms that patients classically identify as more bothersome, involve increased urinary frequency, nocturia, dysuria, and urgency with urge incontinence.[4] Of concern are patients with bladder outlet obstruction (BOO) as this can precipitate a plethora of complications such as secondary urinary tract infections (UTIs), bladder diverticula, and even hydronephrosis or chronic renal failure.

The treatment options for BPH have been subject to dramatic change over the past years with the development of minimally invasive simple prostatectomy and the emergence of advanced medical therapies. In an attempt to reduce perioperative complications associated with open simple prostatectomy (OSP), minimally invasive approaches have evolved. For years, however, OSP surgery remained the procedure of choice despite the alternative of minimally invasive procedures because OSP was believed to be superior concerning long-term functional outcomes. In 2002, Mariano et al. reported their first successful experiences with laparoscopic simple prostatectomy (LSP).[7] Subsequent studies demonstrated that LSP is associated with reduced hospitalization time, reduced morbidity, decreased blood loss, shorter periods of catheterization, and prolonged operative times.[8–10] Despite these improvements LSP has not gained wide acceptance due to technical challenges, restricted range of movements, two-dimensional views, and a steep learning curve.

Robotic-assisted simple prostatectomy (RASP) has evolved to address several of the shortcomings and concerns surrounding LSP. With improved visualization using three-dimensional views and increased degrees of freedom for advanced manipulation and dexterity, RASP holds potential for improving the current standards for simple prostatectomy.[11,12] In 2008 Sotelo et al. first reported RASP in selected patients, demonstrating feasible improvements in perioperative outcomes with an acceptable

A Comprehensive Guide to the Prostate. https://doi.org/10.1016/B978-0-12-811464-3.00015-5

complication rate.[13] Studies have shown that RASP, similar to LSP, is associated with decreased blood loss, shorter catheterization time, and reduced duration of hospital stay compared to OSP.[14–17] In addition, patients who underwent RASP experienced better functional outcomes and a reduction in urinary symptoms as indicated by improvements in International Prostate Symptom Scores (IPSS), reduced postresidual void (PRV), and an overall lower risk of developing incontinence.[18,19] Subsequent publications and emerging evidence suggest that RASP is a safe and feasible treatment option for select patients with large prostates for whom simple prostatectomy is indicated. This chapter will define the indications and contraindications for RASP, describe the three different surgical approaches and explore complications related to the procedure.

Indications and Contraindications

The evaluation of a patient for RASP involves consideration of a variety of factors including the size of the prostate, the extent of urinary symptoms, BPH-associated complications, and patient factors that contraindicate other minimally invasive surgical procedures. Simple prostatectomy is indicated in patients with a large prostate (>80 g) who suffer from moderate to severe lower urinary tract symptoms that are not resolved or alleviated by medical therapy.[20] BPH-related urinary tract symptoms that indicate a need for simple prostatectomy include BOO with urinary retention, recurrent gross hematuria of prostatic origin, recurrent UTIs associated with bladder diverticulum secondary to BOO, bladder calculi secondary to urinary retention and BOO, and pathophysiological changes in the upper tracts such as dilation, hydronephrosis, and chronic kidney failure.[21,22] Patients with BPH and a history of coexisting, surgically correctable conditions such as urinary bladder diverticula, large bladder calculi, and inguinal hernia benefit from RASP as the surgeon can perform a concurrent diverticulectomy, remove bladder calculi, or repair the underlying hernia.[23] Certain patients who are limited in their ability to undergo other minimally invasive surgical procedures such as transurethral resection of the prostate (TURP) may be considered for robotic simple prostatectomy. This population includes patients with a history of hypospadias or urethral stricture diseases for which TURP is contraindicated due to the risk of further trauma to the urethra. Similarly, patients with severe ankylosis of the hip or knee joints or other physically limiting conditions preventing them from assuming lithotomy position required for TURP and other minimally invasive procedures can be considered for RASP instead.

Simple prostatectomy is contraindicated in patients with a carcinoma of the prostate.[24] Patients who have undergone prior pelvic surgery or radiation to the area resulting in obliteration of the surgical planes required to approach the prostate gland should not be considered for simple prostatectomy.[24] Finally, patients with small fibrous prostate glands (under 80 g) should be considered for TURP rather than RASP.

Preoperative Evaluation

Patients are selected as candidates for RASP based on a medical history of symptomatic BPH with bothersome LUTS and a prostate size greater than 80g. Serum prostate-specific antigen (PSA) measurements are obtained and a rectal exam is performed to rule out prostatic carcinoma. Standard preoperative evaluation of patients that qualify as surgical candidates for RASP involves uroflowmetry analysis to determine the peak urinary flow rate (Qmax) and a voiding trial for evaluation of the PRV. Prostatic volume is assessed using transrectal sonographic imaging, and patients are asked to complete the IPSS questionnaire. In addition, medical evaluation and clearance for surgery are required for all patients. Patients are advised to self-administer a lubricant laxative (fleet enema) the day before surgery and should be kept NPO (nothing by mouth) after midnight.

Certain medical conditions or BPH-associated symptoms demand additional preoperative testing and evaluation. Patients with BPH and a history of upper urinary tract pathology including hematuria, recurrent UTIs, or hydronephrosis warrant imaging evaluation with computed tomography (CT) or magnetic resonance imaging (MRI). Evaluation by cystoscopy is indicated for patients with a suspected bladder diverticulum and for patients with a history of urethral stricture diseases or hematuria. Renal function is assessed by blood tests (serum creatinine and blood urea nitrogen) and urine analysis in BPH patients with urinary retention secondary to BOO. Finally, if the patient has an elevated PSA and/or the digital rectal exam is abnormal, a prostate biopsy will be performed in order to rule out the possibility of prostate cancer.

Surgical Technique

Robotic-Assisted Simple Prostatectomy

Compared to LSP, the Da Vinci Surgical System provides improved visualization with a high resolution three-dimensional operative field. The jointed robotic instruments enhance dexterity through seven degrees of freedom allowing the surgeon to complete more challenging and technically demanding techniques during the procedure.[25] Together, these factors have facilitated superior perioperative while demonstrating equivalent improvements regarding long-term functional outcomes compared to OSP. Despite this progress, the main challenge that remains is the demand in technical skill, experience, and the extensive training required by surgeons performing RASP. There are three widely accepted surgical techniques for RASP: the suprapubic, transvesical, and retroperitoneal approach. Currently, there is no evidence that any one approach is superior to the others considering perioperative or long-term functional outcomes. The type of surgical approach is chosen based on the surgeon's preference.[26]

Positioning and Port Placement

Following the induction of general anesthesia, the patient is brought into dorsal lithotomy position in steep Trendelenburg. Legs are secured on stirrups and arms are padded and positioned on the patient's sides. After the patient is secured in the desired position a Foley catheter is placed. Based on preference, the surgeon may decide to perform a cystoscopy and bilateral ureteral stent placement to improve visualization of ureteral orifices during the procedure. Before port-placement is initiated, the stomach is decompressed by means of an orogastric tube.

Once the patient has been appropriately positioned and decompressed, a Veress needle is inserted supraumbilically and used for insufflation of the abdomen. A pneumoperitoneum is established with a set pressure of 15 mmHg. Following successful insufflation, a 12-mm port for the stereoscopic endoscope is placed supraumbilically. Two 8-mm pararectus trocars for the second and third robotic arms are placed laterally on either side of the laparoscope port at a distance of 10 cm. A third 8-mm pararectus trocar is placed in the left lumbar region for the fourth robotic arm. Next, two assistant ports are placed on the right side of the patient to assist in retraction, irrigation, suction, and passing clips as well as sutures. First, a 12-mm air seal trocar is placed 10 cm lateral to the right-sided robotic arm port. Second, a 5-mm assistant port is placed 8 cm superior and 4 cm lateral to the supraumbilical laparoscope port. The da Vinci surgical system robot is placed between the patient's legs and the robotic arms are docked.

Suprapubic Approach

Following the placement of ports, the pelvic space is visualized and assessed with special focus on the presence of adhesions, which may interfere with the surgical procedure. If adhesions are identified in the pelvic space, sharp and blunt dissections are used to improve pelvic accessibility. Once the surgeon has identified the medial umbilical ligaments, the peritoneum is incised on the lateral aspect of both medial ligaments using monopolar electric cautery. Starting at the vas deferens, the incisions are extended in a wide, inverted U-shape fashion through the urachus in the midline.[19] After transecting both medial ligaments and the urachus superior to the bladder neck, any flimsy connective tissue is removed from the space of Retzius using monopolar electric cautery and blunt dissection. Once the bladder is effectively dropped posteriorly after completion of the above-described maneuvers, the anterior surface of the prostate is cleared from fat tissue to improve visualization of its anatomy. The superficial dorsal venous plexus is subsequently secured with cauterization and suture ligation. Two maneuvers can be used to identify the prostatovesical junction prior to bladder neck incision. The Foley catheter can be advanced, which facilitates visualization of the bladder neck. Alternatively, bimanual compression of the bladder neck (bladder neck pinch) using blunt robotic instruments can help localize the prostatovesical junction.[27] Carefully following the appropriate plane of dissection, an incision is made in the midline of the bladder neck using hot shears and Maryland bipolar forceps. After establishing anterior access to the bladder via the bladder neck, the stroma of the prostate is visualized and traction is applied using a 0-silk figure of eight traction suture. Next, the median lobe of the prostate is mobilized and posteriorly exposed by the application of traction using the third robotic arm. Beginning at the posterior border of the bladder neck an incision is made using monopolar electric cautery creating a plane between the transition zone, which is where BPH lesions develop and the peripheral zone that is unaffected by the adenomatous growth. The dissection is extended following the plane between transition and peripheral zones distally until the apex of the prostate is reached. At the apex, the urethra is transected proximal to the external urethral sphincter. Once hemostasis is achieved using electric cautery as well as Hem-o-lok clips (Teleflex Medical), the bladder neck is advanced distally toward the apex of the prostate. Using an absorbable suture a mucosa-to-mucosa anastomosis is created between the bladder neck and the distal segment of the urethra proximal to the external urethral sphincter. Finally, a Foley catheter is carefully advanced through the anterior urethra until it enters the lumen of the bladder while the integrity of the anastomosis is assessed under direct visualization. The temporary application of traction once the Foley catheter is placed and in hemostasis.

Transvesical Approach

The transvesical approach follows the same initial steps described for the suprapubic approach including the establishment of peritoneal access and clearance of the space of Retzius. Following the bladder drop, the dome of the bladder is identified and a vertical midline incision beginning at the dome of the bladder is started to create a midline cystotomy. The transvesical access to the prostate is secured by the placement of four 2-0 Vicryl stay sutures (on a CT-1 needle), which maintain the cystotomy. Following identification and visualization of the prostatic adenoma, a traction suture, delayed absorbable suture is placed through the median lobe of the adenoma to facilitate the dissection.[28] Starting posteriorly, an incision through the bladder mucosa overlying the prostatic adenoma is made. After the plane between the adenoma in the transition zone and the prostatic capsule is identified, enucleation of the prostate is performed using monopolar electric cautery and blunt dissection. The resected adenoma is placed in an Endocatch bag and hemostasis is secured using direct electric cauterization and the placement of ligation sutures at specific bleeding foci in the prostatic fossa.[28] In case of persistent hemorrhage, bleeding is controlled using ligation of the prostatic pedicles using chromic sutures near the bladder neck at the 5-o-clock and 7-o-clock positions. Visualization of the ureteral orifices during the placement of the chromic sutures via administration of intravenous indigo carmine dye can help reduce the risk of incorporating the ureters in the suture ligation of the prostatic pedicles.

After hemostasis is achieved, a 22F three-way Foley catheter with a 30-mL balloon is inserted. The catheter is advanced through the anterior urethra and secured on entering the bladder by inflation of the balloon with 30 mL of sterile water. Closure of the cystotomy is performed in layers using 2-0 V-Loc sutures (Covidien). A Jackson–Pratt drain is placed in the rectovesical pouch, the resected adenoma is extracted, and the fascia and skin are closed. Temporary application of gentle traction to the Foley catheter can facilitate hemostasis.

Retropubic Approach

The retropubic approach follows the same initial steps described for the suprapubic approach including the establishment of peritoneal access and clearance of the space of Retzius. Following the bladder drop, fat is removed from the anterior periprostatic space before the superficial dorsal venous plexus is secured using electric cauterization and suture ligation. A transverse capsular incision is made on the anterior surface of the prostate approximately 2 cm distal to the prostatovesical junction and the prostatic adenoma is identified. Following visualization of the adenoma a plane between the adenoma in the transition zone and the peripheral zone of the prostate is developed. Once the plane is developed, the prostate is enucleated and the adenoma is dissected by blunt dissection using the robotic-curved scissors. Hemostasis is secured with electric cauterization. After complete mobilization of the adenomatous tissue, the urethra is sharply transected proximal to the external urethral sphincter but distal to the seminal colliculus. After completion of the urethral transection under direct visualization, the bladder neck mucosa is advanced distally toward the prostatic apex to approximate the distal urethral mucosa at the site of urethral transection and to achieve retrigonization.[29] Completion of the mucosa-to-mucosa anastomosis between bladder neck and distal urethra is achieved using two figure of eight delayed absorbable sutures or a continuous suture.[26] Closure of the anterior prostatic capsule is completed in watertight manner. Finally, a 20F Foley catheter is inserted and advanced through the anterior urethra and into the bladder. Inflation of the balloon with 40 mL of sterile water secures the Foley catheter and temporary gentle traction can be applied to aid in hemostasis.

Studies have demonstrated the efficacy of RASP in improving perioperative outcomes such as reduced blood loss as well as shorter catheterization and hospitalization times.[8–10] Emerging evidence in the literature reveals that robotic simple prostatectomy also leads to objective improvement in functional outcomes such as urinary symptoms equivalent to those achieved in patients undergoing OSP.[25,26]

A study published by Autorino et al., in 2015 analyzed the experience with RASP across institutions in the United States and across Europe. The median age of the 487 patients was 67 years (range 63–73) with a baseline median IPSS score of 23 and a median maximum flow rate of 8 mL per second.[30] The median PVR (postvoid residual volume) was 108 mL[31] and a median operating time of 154.5 min (range 100–180) and the average amount of blood lost per surgery amounted to 200 mL (range 100–400). Conversion to OSP was reported to be 3% of cases.[30] Zero patients experienced loss of potency or erectile dysfunction following surgery and all experienced significant improvements in IPPS scores and maximal flow rates (see Table 15.1).[30]

Table 15.1 Results of Patients for IPSS, Qmax, and PVR Preoperatively and Postoperatively

	Preoperative		Postoperative	
Parameter	**Median**	**Range**	**Median**	**Range**
IPSS	23	18–27	7	4–9
Qmax (mL/s)	8	5–11	25	20–33
PVR (mL)	108	55–231	Note reported	

IPSS, International Prostate Symptom Scores; *PVR*, postvoid residual volume.

Postoperative Management

The postoperative management of patients who undergo robotic simple prostatectomy mirrors that of most major surgical interventions. Hemoglobin and hematocrit levels are measured on the first postoperative day and patients are encouraged to early ambulation as soon as movement is tolerable. In the first hours following surgery, gentle traction can be placed on the Foley catheter to aid in bleeding control and continuous bladder irrigation is used as needed.[26] The fluid intake and output status is monitored and the output collected by the Jackson–Pratt drain should be recorded. Initially the patient is started on a clear fluid diet, which may be advanced as tolerated. Prior to discharge, patients and their family members are trained in Foley catheter management with leg-bag and the Jackson–Pratt drain is removed. If the patient is tolerating a normal diet, pain control is achieved and there are no abnormalities found on laboratory tests, the patient can be safely discharged home on the first postoperative day. After 1 week, the Foley catheter is removed and PVR is checked to ensure adequate emptying. While the removal of the adenoma leads to significant improvements regarding the urinary function, the risk of developing prostatic carcinoma remains and regional guideline should be followed on screening.

Complications

Early Complications

A wide range of perioperative bleeding complications are reported for robotic simple prostatectomy. A review of robotic simple prostatectomy outcomes evaluated 13 independent series showing a downward trend in blood loss and a reduction in the rate of required transfusions.[26] Among the studies conducted between 2008 and 2015, the mean blood loss has ranged from as low as 50 mL to as high as 558 mL. The largest and most recent series involving 487 patients reported a median blood loss of 200 mL.[30] The rate of required blood transfusions has shown a marked decrease as reported by recent studies.[26] Autorino et al. reported that 1% of patients required blood transfusion due to perioperative blood loss.[30] Delayed bleeding postoperatively is rare. Initial management of delayed bleeding should include bladder irrigation and application of Foley catheter traction.

In addition to perioperative bleeding complications, infections have been described as an early complication following RASP. The most common infections seen in patients include epididymitis, infection of the surgical wound, and UTIs. Surgical wound infections occurred in 0.4% of patients who underwent RASP while reports of wound infections for the open procedure range in prevalence from 2.5% to 4.3%.[30,31] UTIs requiring antibiotic therapy were reported in 4.1% of RASP patients compared to 9.8% of patients who underwent OSP and may be attributed to reduced catheterization times in RASP patients.[9,30] Risk factors for postoperative infections include long-term indwelling Foley catheters, a history of epididymitis and recurrent UTIs.

Incontinence is an uncommon complication. In a large series of 1804 patients subjected to OSP reported postoperative complication rates of 3.7% and 1.2% for early and late urinary incontinence, respectively.[32] These complications are usually attributable to perforation and partial avulsion of the prostatic capsule or avulsion of the urethra at the prostatic apex. Sharp excision of the urethra near the prostatic apex helps in reducing trauma to the external sphincter and leads to improved continence. The prevalence of urge incontinence due to detrusor overactivity in OSP patients was less than 9% and usually transient, resolving within 8 weeks postoperatively. Surgical complication rates are even lower in RASP. In a retrospective review of 487 patients reported that only 2% of patients experienced acute urinary retention requiring catheterization and less than 0.8% of patients suffered from urinary urgency or incontinence.[30]

Late Complications

Bladder neck contracture (BNC) and urethral strictures are rare complication of RASP occurring in only 0.6% of patients who undergo the procedure.[30] The incidence of BNC was reported to be 7.6% in OSP patients with increased frequency among patients who underwent retropubic simple prostatectomy compared to those subjected to the suprapubic approach.[33] BNC is thought to develop from hematomas, urinary leak, or tension at the anastomosis between the bladder neck mucosa and the distal urethral mucosa.[34] Efforts to prevent the incidence of BNC following RASP have focused on ensuring that a watertight anastomosis is created with mucosal apposition and minimal tension. If BNC does ensue, initial management should involve minimally invasive procedures such as urethral dilation or endoscopic incision. Rarely open reconstruction surgery of the bladder neck is required.

Retrograde ejaculation is a common complication of OSP with an incidence similar to that reported for TURP. Recent data suggest that the development of RASP has dramatically increased the preservation of both anterograde ejaculation and erectile function following simple prostatectomy. Among 487 patients in the study conducted by Autorino et al., zero patients experienced complications related to their sexual functioning.[30] In contrast, the rate of erectile dysfunction following OSP was reported to be between 16% and 32% depending on the approach used.[35]

Patients rarely suffer from postoperative paralytic ileus, obstipation, and intolerance of oral intake. The prevalence of ileus was reported to be 0.6% with RASP.[30] Management should include administration of antiemetic therapy and bowel rest. Cardiovascular and thromboembolic events are two of the leading causes of mortality following prostatectomy. Deep vein thrombosis with pulmonary emboli and cardiac failure were reported in 0.6% of patients that underwent robotic simple prostatectomy.[30] Hematuria and clot formation are rare complications and should be managed with bladder irrigation in mild cases or endoscopically as needed.

Summary

RASP is a feasible option for the treatment of patients with clinically significant BPH who are suffering from bothersome LUTS and who have prostate larger than 80 g. Furthermore, RASP provides an alternative to TURP in patients in which TURP is contraindicated. RASP outperforms OSP regarding shorter hospitalization, less blood loss, reduced catheterization time, and overall fewer postoperative complications such as infections or incontinence. In addition, RASP has demonstrated equivalent efficacy compared to OSP in alleviating LUTS while showing superior outcomes regarding the preservation of erectile and sexual function.

References

1. *Epidemiology and pathogenesis of benign prostatic hyperplasia.* UpToDate; 2017.
2. Wei JT, Calhoun E, Jacobsen SJ. Urologic diseases in America project: benign prostatic hyperplasia. *J Urol* April 2005;**173**(4):1256–61.
3. Wei JT, Calhoun E, Jacobsen SJ. Urologic diseases in America project: benign prostatic hyperplasia. *J Urol* May 2008;**179**(5 Suppl.):S75–80.
4. Berry SJ, Coffey DS, Walsh PC, Ewing LL. The development of human benign prostatic hyperplasia with age. *J Urol* September 1984;**132**(3):474–9.
5. Park T, Choi JY. Efficacy and safety of dutasteride for the treatment of symptomatic benign prostatic hyperplasia (BPH): a systematic review and meta-analysis. *World J Urol* August 2014;**32**(4):1093–105.
6. Roehrborn CG. Benign prostatic hyperplasia: an overview. *Rev Urol* 2005;**7**(Suppl. 9):S3–14.
7. Mariano MB, Graziottin TM, Tefilli MV. Laparoscopic prostatectomy with vascular control for benign prostatic hyperplasia. *J Urol* June 2002;**167**(6):2528–9.
8. Porpiglia F, Terrone C, Renard J, et al. Transcapsular adenomectomy(Millin): a comparative study, extraperitoneal laparoscopy versus open surgery. *Eur Urol* January 2006;**49**(1):120–6.
9. McCullough TC, Heldwein FL, Soon SJ, et al. Laparoscopic versus open simple prostatectomy: an evaluation of morbidity. *J Endourol* January 2009;**23**(1):129–33.
10. Baumert H, Ballaro A, Dugardin F, Kaisary AV. Laparoscopic versus open simple prostatectomy: a comparative study. *J Urol* May 2006;**175**(5):1691–4.
11. Gacci M, Bartoletti R, Figlioli S, et al. Urinary symptoms, quality of life and sexual function in patients with benign prostatic hypertrophy before and after prostatectomy: a prospective study. *BJU Int* February 2003;**91**(3):196–200.
12. Varkarakis I, Kyriakakis Z, Delis A, Protogerou V, Deliveliotis C. Long-term results of open transvesical prostatectomy from a contemporary series of patients. *Urology* August 2004;**64**(2):306–10.
13. Sotelo R, Clavijo R, Carmona O, et al. Robotic simple prostatectomy. *J Urol* February 2008;**179**(2):513–5.
14. John H, Bucher C, Engel N, Fischer B, Fehr JL. Preperitoneal robotic prostate adenomectomy. *Urology* April 2009;**73**(4):811–5.
15. Matei DV, Brescia A, Mazzoleni F, et al. Robot-assisted simple prostatectomy (RASP): does it make sense? *BJU Int* December 2012;**110**(11 Pt C):E972–9.
16. Vora A, Mittal S, Hwang J, Bandi G. Robot-assisted simple prostatectomy: multi-institutional outcomes for glands larger than 100 grams. *J Endourol* May 2012;**26**(5):499–502.
17. Yuh B, Laungani R, Perlmutter A, et al. Robot-assisted Millin's retropubic prostatectomy: case series. *Can J Urol* June 2008;**15**(3):4101–5.
18. Banapour P, Patel N, Kane CJ, Cohen SA, Parsons JK. Robotic-assisted simple prostatectomy: a systematic review and report of a single institution case series. *Prostate Cancer Prostatic Dis* March 2014;**17**(1):1–5.

19. Robotic assisted laparoscopic simple suprapubic prostatectomy – the Smith Institute for urology experience with an evolving technique. October 2014;**1**(1):55–9.
20. McVary KT, Roehrborn CG, Avins AL, et al. Update on AUA guideline on the management of benign prostatic hyperplasia. *J Urol* May 2011;**185**(5):1793–803.
21. Patel MN, Hemal AK. Robot-assisted laparoscopic simple anatomic prostatectomy. *Urol Clin North Am* November 2014;**41**(4):485–92.
22. Sosnowski R, Borkowski T, Chlosta P, et al. Endoscopic simple prostatectomy. *Cent Eur J Urol* 2014;**67**(4):377–84.
23. Raheem O, Parsons K. *Is there still a role for simple prostatectomy today*. 2016.
24. Sutherland DE, Perez DS, Weeks DC. Robot-assisted simple prostatectomy for severe benign prostatic hyperplasia. *J Endourol* April 2011;**25**(4):641–4.
25. Finkelstein J, Eckersberger E, Sadri H, Taneja SS, Lepor H, Djavan B. Open versus laparoscopic versus robot-assisted laparoscopic prostatectomy: the European and US experience. *Rev Urol* Winter 2010;**12**(1):35–43.
26. Holden M, Parsons JK. Robotic-assisted simple prostatectomy: an overview. *Urol Clin North Am* August 2016;**43**(3):385–91.
27. Ost MC, In: Ost MC, editor. *Robotic and laparoscopic reconstructive surgery in children*. Springer; 2011.
28. Leslie S, Abreu AL, Chopra S, et al. Transvesical robotic simple prostatectomy: initial clinical experience. *Eur Urol* August 2014;**66**(2):321–9.
29. Wang S, Yang C, Chang C. Robotic simple prostatectomy: initial single-center experience in Taiwan. June 2016;**27**(2):77–80.
30. Autorino R, Zargar H, Mariano MB, et al. Perioperative outcomes of robotic and laparoscopic simple prostatectomy: a European-American multi-institutional analysis. *Eur Urol* July 2015;**68**(1):86–94.
31. Lucca I, Shariat SF, Hofbauer SL, Klatte T. Outcomes of minimally invasive simple prostatectomy for benign prostatic hyperplasia: a systematic review and meta-analysis. *World J Urol* April 2014;**33**(4):563–70.
32. Serretta V, Morgia G, Fondacaro L, et al. Open prostatectomy for benign prostatic enlargement in southern Europe in the late 1990s: a contemporary series of 1800 interventions. *Urology* October 2002;**60**(4):623–7.
33. Roos NP, Ramsey EW. A population-based study of prostatectomy: outcomes associated with differing surgical approaches. *J Urol* June 1987;**137**(6):1184–8.
34. Simhan J, Ramirez D, Hudak SJ, Morey AF. Bladder neck contracture. *Transl Androl Urol* June 2014;**3**(2):214–20.
35. Benign prostatic hyperplasia: diagnosis and treatment. Guideline overview. Agency for Health Care Policy and Research. *J Natl Med Assoc* July 1994;**86**(7):489, 548–9.

Transurethral Microwave Thermotherapy for Treatment of Benign Prostatic Hyperplasia

16

Nathaniel D. Kwok, Dominique Thomas, Bilal Chughtai
Weill Cornell Medicine, New York, NY, United States

Introduction

When treating symptomatic benign prostatic hyperplasia (BPH), conventional pharmacological and/or surgical interventions can be either contraindicated, poorly tolerated, or simply ineffective in producing therapeutic efficacy. The correct choice of a minimally invasive therapy treatment is based on relative efficacy, risk, availability, and expertise on the part of the care provider.

Transurethral microwave thermotherapy (TUMT) is an office-based procedure that uses heat to ablate hyperplastic prostate tissue. This is achieved by first positioning a microwave antenna inside the prostate (via a specially designed urinary catheter) to target the area of urethral obstruction, and then delivering microwave energy that heats prostate tissue to a necrotic range of 45–55°C.

Although the notion of microwave therapy in the treatment of BPH stems back to the 1980s, it is only within the last decade or so that TUMT has begun to emerge as an attractive alternative to standard BPH therapies, including transurethral resection of the prostate (TURP, the current "gold-standard" therapy) and medication (namely alpha-blockers and 5-alpha reductase inhibitors)—owing to the delivery to precise, high-energy microwave radiation to the target area without harming nearby structures.

TUMT is a safe, less invasive, and comparably effective alternative to standard surgical therapy. In the American Urological Associations (AUAs) 2011 update on BPH management, TUMT was deemed "effective in partially relieving LUTS [lower urinary tract symptoms] secondary to BPH and may be considered in men with moderate or severe symptoms."[1] A 2012 systematic review of 15 randomized controlled trials found the International Prostate Symptom Score (IPSS) to be reduced by 65%, and peak urinary flow rate to be increased by 79% in TUMT-treated patients after 12 months, on average (albeit compared to 77% and 119%, respectively, in TURP-treated patients).[2] The same review found TUMT-treated patients to have significantly fewer adverse posttreatment events—including blood transfusion due to blood loss, strictures, and sexual dysfunction—than those treated with TURP.

A Comprehensive Guide to the Prostate. https://doi.org/10.1016/B978-0-12-811464-3.00016-7

Historical Context

The use of heat, for resolving BOO from BPH has long been established. In 1921, McCaskey found that delivering repeated doses of quartz ultraviolet light to the prostate, transrectally, successfully alleviated prostatic obstruction in "45 cases of 'the types that commonly harass the physician to the desperation point,'"[3] with rectal bleeding as a notable side effect (stemming from repeated rectal probing). Corbus, in 1929, used a diathermic rectal probe heated to 108–122°F—just below patients' pain tolerance threshold—to deliver treatment in a single 40-min session. With this technique, Corbus found prostatic obstruction to have subsided (within 10 days of treatment) in 15 out of 17 patients (88%), with 1 patient reporting a major adverse effect (a developing abscess that ruptured on treatment).[4]

Despite these promising results, heat therapy for BPH would not gain significant clinical traction for several decades thereafter. This was due in large part to the then-ongoing emergence of TURP as the dominant surgical intervention for BPH—a development made possible by the invention of the "Stern-McCarthy resectoscope" (1931), a tool allowing for precise cutting and coagulation of prostate tissue[5]—the unrivaled gold standard for the next 70 years. To this day, TURP remains the most utilized surgical procedure for BPH, but several competing therapies such as TUMT are now quickly closing that gap.

Early evidence suggested increased heat sensitivity among neoplastic cells, relative to normal cells. Reignited interest in heat as a potential ablative therapy for BPH, particularly among patients for whom invasive surgery (such as TURP) was contraindicated. Yerushalmi et al. (in 1982, 1985) demonstrated that transrectal microwave therapy was efficacious in achieving a "hyperthermic" desired temperature range of 42–45°C, which was thought to be high enough to effect coagulative necrosis in neoplastic cells, while preserving normal prostate tissue.[6] Subsequent comparisons of transrectal versus transurethral hyperthermia found the latter to be more efficacious[7] (an unsurprising finding given BPHs predominance in the prostatic transition zone, nearest to the urethra); transurethral methods have since predominated.

It was apparent that another technological breakthrough was needed, if microwave therapy was to become a viable treatment for BPH. Firstly, temperatures of 42–45°C were found to be incapable of consistently inducing coagulative necrosis in prostatic tissue; in fact, observable histologic change of any kind (e.g., protein denaturation, prostatic cell death) was generally absent at those temperatures.[8] Subsequent experiments corroborated this finding, further noting that symptomatic relief was also observed in patients for whom hyperthermia failed to induce necrosis but succeeded in denervating nearby urethral and periurethral prostatic tissue, suggestive of an unintended alternate mechanism of action.[9] Finally, the urethral pain tolerance threshold was found to be 45°C,[10] eliminating the option of simply raising the ablation temperature. This threshold has come under some scrutiny.[11]

The development of urethral cooling mechanisms, or "thermotherapy" (so named to distinguish it from its technological predecessor, hyperthermia), defined simply by intraprostatic ablation temperatures of greater than 45°C.

Modern TUMT (or "thermotherapy," so named to distinguish it from its technological predecessor, hyperthermia) achieves ablation temperatures of greater than 45°C (with some reaching 70°C or higher) while keeping urethral temperature below its pain threshold.

Clinical Indications and Procedure

Indications

The AUA considers TUMT to be an effective option for treating men with moderate or severe LUTS secondary to BPH.[1] As with the other surgical BPH interventions, TUMT is indicated only when there is evidence of bladder outlet obstruction, and additionally only after medical therapy (which include alpha-blockers, 5-alpha reductase inhibitors, and anticholinergics) has proven to be ineffective, poorly tolerated, or refused. When considering TUMT among other interventions (which includes the gold standard, TURP), among the things to consider are the patient's preferred setting (hospital vs. outpatient), degree of outlet obstruction, and relative risk for events such as stricture, blood loss, and incontinence, as well as other complications associated with surgery and/or anesthesia.

Contraindications

Standard protocol for transurethral procedures apply: sterile urine culture must be recorded, and preoperative evaluation for prostate and urothelial carcinoma must be conducted. The latter is particularly important because, unlike TURPs, TUMTs procedure does not produce a tissue sample.

Patients with atypical pelvic anatomy, including those with prior history of TURP or other regional trauma, are not suitable candidates for TUMT. Patients whose prostate measures under 25 g or over 100 g, whose prostatic urethra is less than 3 cm in length, or who have an enlarged or prominent median lobe do not typically respond well to TUMT; for this reason, preoperative transrectal ultrasound is recommended. Implanted pacemakers or defibrillators are contraindications unless cleared for shutoff during the procedure.

Remaining contraindications include penile prosthesis, urinary sphincters, urethral stricture, peripheral arterial disease with intermittent claudication or Leriche's syndrome, prostatic ball-valve median lobe, and evidence of prostate or bladder cancer.[11] Hip replacement is neither longer considered a contraindication nor is acute urinary retention.[10]

Preprocedure

Prophylactic antibiotics are to be administered to catheterized patients and to those with recently manipulated urinary tracts. Oral analgesics or anxiolytics can be given as needed as per the physician's discretion.

After patients void their bladder to the best of their ability, they are then catheterized to drain any remaining urine in the bladder, replacing with 40 cc of distilled water. Then the urethra is locally anesthetized, and treatment catheter is inserted and positioned within the prostatic urethra, inflating the securing balloon once correct positioning has been verified. Lastly, a temperature probe is often inserted into the rectum.

Postprocedure

The immediate effect of microwave therapy is prostatic edema, which may obstruct the urethra and cause urinary retention, so a urinary catheter is inserted for several days following the procedure. The length of time for which the catheter must remain in place is usually 3–5 days but is up to the discretion of the treating physician and other factors such as preoperative urinary retention and prostate size.

The Biophysics and Technical Features of Modern Transurethral Microwave Thermotherapy

The Microwave Antenna Catheter

TUMT is performed via a urinary catheter that is inserted into the urinary meatus of the penis and threaded proximally to inflate within the bladder, which securely positions the specialized segment of the catheter against the urethral surface of the prostate transition zone. This specialized segment is outfitted with an antenna that produces microwave energy at frequencies of 0.9–1.3 MHz, allowing the microwaves to penetrate roughly 15 mm of water-rich tissue.[11] Microwave energy causes electrical dipoles and charge carriers (chiefly water molecules and free ions, respectively) to oscillate, giving off thermal energy that heats the surrounding tissue to levels sufficient to induce protein denaturation and coagulative necrosis.

Inducing Necrosis via Heat

Coagulative necrosis, occurs when the thermal energy delivered to tissue reaches a sufficient threshold. Thermal dosage can be accurately modeled as a function of two parameters: temperature and heat duration. Relating these two quantities in a single biparametric function gives rise to the notion of "thermal *isoeffect* dose," which is conventionally expressed in "cumulative equivalent minutes at 43°C": $CEM_{43} = \Delta t \cdot R^{(43-T)}$, where Δt is the length of heat exposure in minutes, T is the average temperature over t, and R is a constant equal to 0.25 for $T < 43°C$ and 0.5 for $T > 43°C$.[12] This measure—derived from the Arrhenius model of thermal damage (from which an exponential function reduces to a constant approximation, R, with high accuracy between 41 and 57°C)[13,14]—allows for a reasonable estimation of the expected necrosis induced by various hyperthermic and thermotherapeutic methods varying in both time and temperature.

The CEM_{43} model makes apparent the appeal of higher thermotherapeutic temperatures, as the exponential base coefficient R can be interpreted to mean that, for each temperature increase of 1°C, equivalent necrosis should be achieved in one-quarter the time (if <43°C) or one-half the time (if >43°C). If increasing T from 45 to 47°C decreases the required Δt from 60 to 15 min, then it can be certain that TUMT can be performed within a single office visit.

In canine model, prostate tissue damage at a $CEM_{43}=80$ min and complete coagulative necrosis at $CEM_{43}=240$ min.[11] Extrapolated to human tissue, this finding can be used to explain why early attempts at coagulative hyperthermia—which was administered at 42.5–43°C, over 1 h—repeatedly failed to demonstrate the intended effects on histological examination. By contrast, TUMT exceed this threshold.

Intraprostatic Temperature Monitoring

Drawing analogy to the Pennes equation of in vivo heat transfer,[6] there are three factors that collectively determine the achieved temperature in a patient's prostate tissue. Two of those three factors—the specifications of the microwave generator (representing the "metabolic" heat source in Pennes), and the heat conductivity of prostate tissue—vary either insignificantly or not at all from patient to patient. But the third influencing factor, blood flow (which causes heat loss), varies widely, both across different patients as well as within the same patient. In the context of thermal stress (such as that induced by TUMT), variability in patients' sympathetic nervous system responses further contributes to the unpredictability of perfusion levels, which consequently makes it extremely difficult to predict intraprostatic temperatures (and therefore tissue damage), if all that is known are the specifications of the microwave power source.[15]

Following this, Bolmsjö et al., with a simulated TUMT model, found that realistic differences in instraprostatic perfusion (which has been reported to as much as double during TUMT) would result in extreme variability in extent of coagulated tissue for a given specification of power and duration.[16] This and similar results served to motivate development of a modality (ProstaLund's CoreTherm aka *ProstaLund Feedback Treatment*, henceforth abbreviated as CoreTherm/PLFT) that continually monitors intraprostatic temperatures in three different locations and modulates power delivery accordingly to deliver precise and individualized therapy.[17] CoreTherm/PLFT is the first (and remains the only) device that can actively monitor and control its thermal damage (as this calculation requires knowing the temperature of the microwave-absorbing tissue)—to date there is yet to exist any conclusive clinical evidence in favor of one particular TUMT modality over any other.[1]

Urethral Cooling

Another feature of some modern TUMT devices is a urethral cooling mechanism, generally involving the circulation of cooling fluid (typically water) within the catheter's microwave-emitting zone. This is intended both to protect the prostatic urethra from the necrotic damage and sensation of discomfort/pain that is thought to occur at temperatures required to achieve coagulative necrosis. Variations of this feature are present in all three successive generations of Urologix devices (Prostatron, Targis, and Cooled ThermCath, respectively) as well as in Medifocus' Prolieve, the former of that has been particularly vigilant about maximizing the cooling capabilities of its devices. The efficacy and safety profiles of TUMT devices employing cooled thermotherapy have largely been positive, faring comparably to other third-generation devices (including noncooling ones), as noted in the previous paragraph. Yet, that the efficacy and safety profiles of these devices are not decidedly superior to their noncooling counterparts' calls into question the purported indispensability of this feature. A 1-year posttreatment analysis of Targis on patients with chronic prostatitis found no significant differences, between cohorts treated at either 55 or 70°C, across various measures of efficacy, tolerability, and adverse reactions[18]; this finding should serve to motivate future work, in BPH patients, which investigates the supposed therapeutic benefit of supracoagulative temperatures (>60°C, where urethral cooling is clearly necessary). Also notable is that 55°C is roughly the operating temperature of both of the noncooling TUMT devices (CoreTherm/PLFT and TherMatrx TMx-2000), which have not been associated with intolerance or urethral necrosis. Due to findings such as these, some have speculated that the prior association of early generation thermotherapy devices with urethral involvement was attributable more to poor, heat-absorbing antenna design than to the microwave radiation itself.

In short, the six commercially available, FDA-approved TUMT devices—CoreTherm/PLFT, Prostatron, Targis, Cooled ThermCath/CCT, Prolieve, and TMx-2000—have all been found, largely, to be effective and safe in published literature, which nonetheless remains conspicuously sparse, making it difficult to draw definitive conclusions in favor of any particular device over another. Additionally, further critical examination on specific design features should be done to help separate veracity from branding. Such features include power delivery, therapeutic temperature (thermotherapeutic vs. ablative range), instaprostatic temperature monitoring and feedback, urethral cooling, and antenna design.

TUMT Versus TURP

TURP has long been the definitive surgical intervention for benign BPH-induced, moderate to severe LUTS. TURPs well-established efficacy, however, has always been hampered by a conspicuously high rate of complications, with total short and long-term morbidity at one point reported to exceed 20%.[1] As a minimally invasive procedure, the promise of TUMT is to minimize risk of complication while maintaining an acceptable level of therapeutic efficacy (this latter expectation has risen dramatically with the introduction of higher-energy TUMT modalities).

Efficacy

In a systematic review conducted by the Cochrane Urology Group found TUMT to be less efficacious than TURP (pooling data from six nonblinded randomized controlled trials) with respect to LUTS reduction, peak urinary flow (Qmax) improvement, and retreatment rate, and more efficacious than a sham control treatment (in eight subject-blinded RCTs) by the same measures. Below is a summary and discussion of these results, which these authors consider to represent the highest quality data available in the published literature.

In the six RCTs comparing TUMT to TURP included in the systematic review,[19–24] **pooled mean symptom scores** at 12-month follow-up decreased by 65% in the combined TUMT cohort (from 19.4 pretreatment to 6.7 posttreatment) versus 77% in the combined TURP cohort (from 19.6 to 4.5), indicating that TURP was more effective in this regard. Note that Madsen–Iversen scores were normalized to the IPSS scale to compute the pooled mean. The weighted mean differences (WMDs) in the posttreatment Madsen–Iversen and IPSS scores between cohorts were −1.59 ($P<.05$) and −1.00 ($P=.05$), respectively, favoring TURP. Of the two long-term randomized trials, Wagrell et al.[25], using the CoreTherm/PLFT modality, found the differences in mean symptom score to hold consistent over time (8.2 vs. 5.0 at 36 months, 7.4 vs. 6.0 at 60 months),[25] whereas Floratos et al. (using Prostatron 2.5) found them to diverge further in favor of TURP (12 vs. 3 at 36 months).[22] Qualitatively, both treatments consistently produced 12-month follow-up mean symptom scores in the "mild" range; their mean pretreatment scores were on the borderline of "moderate" and "severe."

With respect to the 12-month Madsen–Iversen and IPSS WMDs (−1.59, −1.00)—it's unclear whether a statistical difference of this magnitude translates to clinical significance. A **posttreatment quality-of-life** (QoL) survey given in three RCTs[22–24]—where patients rated their perception of present urinary condition from 1 to 5 (lower is better)—found no statistically significant QoL difference between treatment cohorts (1.6 [TUMT] vs. 1.2 [TURP]). Contrast this with data from six TUMT versus sham control RCTs,[26–31] which measured posttreatment Madsen–Iversen and IPSS score WMDs of −5.10 ($P<.001$) and −5.15 ($P<.001$), strongly favoring TUMT over the sham control, along with a similarly, significant discrepancy in mean QoL score (1.6 vs. 3.1, favoring TUMT).

Peak urinary flow (or *Qmax*) improved, at 12 months, on average by 70% (from 7.9 to 13.5 mL/s) in the TUMT cohort versus 119% (from 8.6 to 18.7 mL/s) in the TURP cohort, with a WMD of 5.08 mL/s (95% CI 3.88 to 6.88) favoring TURP. Of interest is that all five RCTs involving Prostatron devices (which include both the high energy 2.5 and lower energy 2.0 iterations) recorded statistically significant WMDs favoring TURP, whereas the lone RCT (Dahlstrand et al.) involving CoreTherm/PLFT did not record a significant WMD.[20] Looking at the two longitudinal RCTs, Wagrell et al.[25] found no significant difference relative to TURP in mean peak urinary flow at 36 months (11.9 [TUMT] vs. 13.5 mL/s [TURP], $P=.58$) or 60 months (11.4 vs. 13.6 mL/s),[25] while Floratos et al. found TURPs advantage to widen at 36 months (11.9 vs. 24.7 mL/s, $P<.01$).[22] Note that the between-study disparity at 36 months is actually on the TURP side (and thus should not be attributed to TUMT modal differences); with more than 50% 36-month attrition in one study, this is possibly an artifact of small sample size. Encouragingly, mean peak flow for the CoreTherm/PLFT (TUMT) cohort largely kept pace with TURP over the 60-month duration and declined by only 14% within that timeframe. TUMT versus sham control RCTs found peak flow rates improved to a significantly greater extent in the TUMT cohort (43%) than in the sham control cohort (11%) after 3–6 months, which is unsurprising.

Retreatment for BPH symptoms (with either medication or secondary procedure) was indicated for every 7.54/100 person-years in the TUMT cohort versus 1.05/100 person-years in the TURP cohort (HR = 10.0, 95%CI 2.44–50). When comparing TUMT to sham control, retreatment was indicated every 1.5/100 person-years for TUMT and 13.5/100 person-years for the sham control (HR = 0.12, 95%CI 0.03–0.48). However, none of the RCTs included in the systematic review collected retreatment data for longer than 12 months. In an RCT excluded from the review (excluded because the study was primarily concerned with economic analysis), De La Rosette et al.[32] followed 144 patients randomized to either Prostatron 2.5 HE-TUMT or TURP for 3 years, finding their respective 3-year retreatment risk to be 22.9% (8.62/100 person-years) and 13.2% (4.71/100 person-years).[32] Although this intermediate-term finding did not reach statistical significance ($P = .21$), it is nonetheless consistent with the trend of TURP proving somewhat more durable than TUMT by this measure.

Complications

In the systematic review comparing TUMT to TURP, TUMT participants experienced significantly fewer of the following events: blood transfusion (0.0% vs. 12%; RR = 0.11, 95%CI 0.01–0.86), clinically important hematuria (1.2% vs. 7.2%; RR = 0.25, 0.07–0.85), transurethral resection syndrome (0.0% vs. 6.1%; RR = 0.13, 0.02–0.81), retrograde ejaculation (22% vs. 58%; RR = 0.39, 0.21–0.75), and erectile dysfunction (5.7% vs. 14%; RR = 0.41, 0.16–1.05). Similarly, strictures of the meatus, urethra, or bladder neck occurred far more often within the TURP cohort, requiring surgical intervention at a rate of 5.85/100 person-years versus 0.63/100 person-years in the TUMT cohort (HR = 9.76, $P < .001$). On the other hand, TUMT participants developed significantly more dysuria/urgency (31% vs. 13%; RR = 2.22, 1.28–3.86) and urinary retention (24% vs. 6.9%; RR = 2.94, 1.52–5.70) than TURP participants. No deaths were recorded in any case.

In comparisons to sham control, TUMT participants developed significantly more urinary retention, (self-limited) dysuria, and posttreatment hematuria, but not-significantly more strictures, urinary tract infections, urinary incontinence, and retrograde ejaculation. Again, there were no recorded deaths.

Clinical Implications and Conclusion

As a method, TUMT upholds its promise of being a relatively safe and effective alternative to TURP for treating moderate-to-severe symptoms of BPH, with the decided advantages of being minimally invasive and outpatient. On average, modern TUMT modalities have proven to be somewhat less efficacious than TURP at lowering subjective symptom scores and increasing Qmax (peak urinary flow rate), which is offset by a lower incidence of serious perioperative complications including stricture, transfusion-requiring blood loss, transurethral resection syndrome, and sexual dysfunction. At the same time, TUMT appears to be associated with higher rates of symptom-driven retreatment and development of postoperative dysuria and urine retention (though these are generally self-limiting). Accordingly, a good candidate for TUMT might be someone whose Qmax is not critically compromised, is sexually active, and who might be at additional risk for perioperative blood loss or complications from anesthesia. Conversely, someone with postrenal acute kidney injury might want to stick with TURP for its more pronounced Qmax improvement and lower rates of symptomatic reoccurrence. Finally, it should be recognized that the randomized controlled trials compiled in the 2012 Cochrane systematic review, constituting our core basis of high-quality data, nonetheless vary considerably in modality of TUMT, sample size, measured outcomes, and follow-up attrition, among other things—to give a sense of the bounds on its validity.[2]

The published literature serves reasonably well to inform patients and clinicians about the benefits and risks that can be expected of TUMT as a general method, yet it remains conspicuously sparse in controlled trials comparing the various specific modalities that modern-generation TUMT comprises. Indeed, with such differentiating intermodality features as thermocoagulation versus thermoablation, continuous intraprostatic feedback regulation, urethral cooling, and intraprostatic adrenaline injections,[33] there now exists about as much technological variety within TUMT as there is between TUMT and other classes of surgical interventions for BPH (which is not to say we do not also need more quality data comparing TUMT to other techniques—we do). More granular future research into the veracity of specific innovations will allow for providers to base their choice of device more so on science than on marketing. Granularity is trickier than ever to achieve amid today's climate of unrelenting technological inundation—within BPH alone, there are at least four other families of semi- or minimally invasive techniques that have gained traction in recent years (HIFU, HoLEP, TUVP, TUNA, among others)—nonetheless, it remains no less necessary.

References

1. McVary KT, Roehrborn CG, Avins AL, et al. Update on AUA guideline on the management of benign prostatic hyperplasia. *J Urol* 2011;**185**(5):1793–803.
2. Hoffman RM, Monga M, Elliott SP, et al. Microwave thermotherapy for benign prostatic hyperplasia. *Cochrane Database Syst Rev* 2012;(9).
3. McCaskey D. Quartz light therapy in obstinate prostatic affections. *Int Med Surg Surv Obstet Pediatr* 1921;**2**(6):2.
4. Edwards L. History of nonsurgical treatment. In: Frank Hinman J, editor. *Benign prostatic hypertrophy*. New York: Springer Science & Business Media; 2012. p. 30–4.
5. Herr HW. The enlarged prostate: a brief history of its surgical treatment. *BJU Int* 2006;**98**(5):947–52.
6. Yerushalmi A, Servadio C, Leib Z, Fishelovitz Y, Rokowsky E, Stein JA. Local hyperthermia for treatment of carcinoma of the prostate: a preliminary report. *Prostate* 1982;**3**(6):623–30.

7. Stawarz B, Szmigielski S, Ogrodnik J, Astrahan M, Petrovich Z. A comparison of transurethral and transrectal microwave hyperthermia in poor surgical risk benign prostatic hyperplasia patients. *J Urol* 1991;**146**(2):353–7.
8. Eisenberg ER, Badlani GH. Long-term treatment outcomes of CoreTherm microwave feedback thermotherapy. *Curr Urol Rep* 2004;**5**(4):287–94.
9. Denervation of periurethral prostatic tissue by transurethral microwave thermotherapy. *Scand J Urol Nephrol* 2000;**34**(1):42–5.
10. Rubeinstein JN, McVary KT. Transurethral microwave thermotherapy for benign prostatic hyperplasia. *Int Braz J Urol* 2003;**29**:251–63.
11. Walmsley K, Kaplan SA. Transurethral microwave thermotherapy for benign prostate hyperplasia: separating truth from marketing hype. *J Urol* 2004;**172**(4):1249–55.
12. Yarmolenko PS, Moon EJ, Landon C, et al. Thresholds for thermal damage to normal tissues: an update. *Int J Hyperth* 2011;**27**(4):320–43.
13. Pearce JA. Relationship between Arrhenius models of thermal damage and the CEM 43 thermal dose. 2009.
14. Dewey WC. Arrhenius relationships from the molecule and cell to the clinic. *Int J Hyperth* 2009;**25**(1):3–20.
15. Berges R. TUMT for benign prostatic hyperplasia. In: Chaussy C, editor. *Therapeutic energy application in urology: standards and recent developments*. Georg Thieme Verlag; 2005.
16. Bolmsjö M, Sturesson C, Wagrell L, Andersson-Engels S, Mattiasson A. Optimizing transurethral microwave thermotherapy: a model for studying power, blood flow, temperature variations and tissue destruction. *Br J Urol* 1998;**81**(6):811.
17. Larson BT, Bolmsj MB, Wagrell L, Larson TR. Prostalund feedback thermotherapy: a review. *Curr Urol Rep* 2003;**4**(4):292–6.
18. Kastner C, Hochreiter W, Huidobro C, Cabezas J, Miller P. Cooled transurethral microwave thermotherapy for intractable chronic prostatitis—results of a pilot study after 1 year. *Urology* 2004;**64**(6):1149–54.
19. Ahmed M, Bell T, Lawrence WT, Ward JP, Watson GM. Transurethral microwave thermotherapy (Prostatron version 2.5) compared with transurethral resection of the prostate for the treatment of benign prostatic hyperplasia: a randomized, controlled, parallel study. *Br J Urol* 1997;**79**(2):181–5.
20. Dahlstrand C, WaldÉN M, Geirsson G, Pettersson S. Transurethral microwave thermotherapy versus transurethral resection for symptomatic benign prostatic obstruction: a prospective randomized study with a 2-year follow-up. *Br J Urol* 1995;**76**(5):614–8.
21. D'ancona F, Witjes WP, Welling L, Debruyne FM, De La Rosette JJ. Transurethral resection of the prostate vs high-energy thermotherapy of the prostate in patients with benign prostatic hyperplasia: long-term results. *Br J Urol* 1998;**81**(2):259–64.
22. Floratos DL, Kiemeney LALM, Rossi C, Kortmann BBM, Debruyne FMJ, de la Rosette JJMCH. Long-term Followup of randomized transurethral microwave thermotherapy versus transurethral prostatic resection study. *J Urol* 2001;**165**(5):1533–8.
23. Nørby B, Nielsen HV, Frimodt-Møller PC. Transurethral interstitial laser coagulation of the prostate and transurethral microwave thermotherapy vs transurethral resection or incision of the prostate: results of a randomized, controlled study in patients with symptomatic benign prostatic hyperplasia. *BJU Int* 2002;**90**(9):853–62.
24. Wagrell L, Schelin S, Nordling J, et al. Feedback microwave thermotherapy versus TURP for clinical BPH—a randomized controlled multicenter study. *Urology* 2002;**60**(2):292–9.
25. Wagrell L, Schelin S, Nordling J, et al. Three-year follow-up of feedback microwave thermotherapy versus TURP for clinical BPH: a prospective randomized multicenter study. *Urology* 2004;**64**(4):698–702.

26. Bdesha AS, Bunce CJ, Snell ME, Witherow RO. A sham controlled trial of transurethral microwave therapy with subsequent treatment of the control group. *J Urol* 1994;**152**(2 Pt 1):453–8.
27. Blute ML, Patterson DE, Segura JW, Tomera KM, Hellerstein DK. Transurethral microwave thermotherapy v sham treatment: double-blind randomized study. *J Endourol* 1996;**10**(6):565–73.
28. De Wildt MJ, Hubregtse M, Ogden C, Carter SS, Debruyne FM, De la Rosette JJ. A 12-month study of the placebo effect in transurethral microwave thermotherapy. *Br J Urol* 1996;**77**(2):221–7.
29. Larson TR, Blute ML, Bruskewitz RC, Mayer RD, Ugarte RR, Utz WJ. A high-efficiency microwave thermoablation system for the treatment of benign prostatic hyperplasia: results of a randomized, sham-controlled, prospective, double-blind, multicenter clinical trial. *Urology* 1998;**51**(5):731–42.
30. Nawrocki JD, Bell TJ, Lawrence WT, Ward JP. A randomized controlled trial of transurethral microwave thermotherapy. *Br J Urol* 1997;**79**(3):389–93.
31. Roehrborn CG, Preminger G, Newhall P, et al. Microwave thermotherapy for benign prostatic hyperplasia with the dornier urowave: results of a randomized, double-blind, multicenter, sham-controlled trial. *Urology* 1998;**51**(1):19–28.
32. De La Rosette JJMCH, Floratos DL, Severens JL, Kiemeney LALM, Debruyne FMJ, Pilar Laguna M. Transurethral resection vs microwave thermotherapy of the prostate: a cost-consequences analysis. *BJU Int* 2003;**92**(7):713–8.
33. Knutson T, Johansson A, Damber JE, Fall M, Vesely S, Peeker R. Intraurethral prostate injections with mepivacaine epinephrine: effects on patient comfort, treatment time and energy consumption during high-energy transurethral microwave thermotherapy. *Scand J Urol Nephrol* 2009;**43**(4):300–6.

Transurethral Needle Ablation of the Prostate

17

Caroline Chung, Dominique Thomas, Bilal Chughtai
Weill Cornell Medicine, New York, NY, United States

Introduction

Benign Prostatic Hyperplasia (BPH) is a noncancerous enlargement of the prostate gland that can lead to lower urinary tract symptoms (LUTS), previously referred to as prostatism. LUTS due to bladder outlet obstruction from BPH can include frequency, slow urinary stream, nocturia, and other disruptive symptoms, which detrimentally affect quality of life for almost half of men in the United States by the time they are 80 years old.[1]

The progression of LUTS to more serious genitourinary conditions such as acute urinary retention (AUR) is associated with BPH. Medical and surgical management of patients with mild to moderate BPH suffering from LUTS has been shown to aid in the prevention of AUR.[2] Untreated BPH can also lead to recurrent urinary tract infections (UTIs) and bladder calculi. First-line treatment currently consists of α-receptor blockers and/or 5α-reductase inhibitors.[3]

In patients for whom starting or continuing medications is contraindicated due to either intolerance or refractory disease. Transurethral needle ablation of the prostate (TUNA) is a minimally invasive procedure, which can improve or eliminate LUTS due to BPH. TUNA presents several advantages over other BPH treatment modalities such as transurethral resection of the prostate (TURP), which is associated with significant rates of retrograde ejaculation. One of TUNA's major strengths is that it is minimally invasive and can be performed under local anesthesia in the office. TUNA allows for an incremental ablation of the prostate tissue, which offers precision over more invasive techniques.

Transurethral Needle Ablation of the Prostate

Nonmedical BPH treatments have been developed to meet the needs of patients who are unable to take or remain on medications. For those who are not candidates for invasive surgical approaches, TUNA is a practical alternative that has the advantage of being minimally invasive, cost-effective, and efficacious. TUNA falls under the umbrella of electrosurgical, noncauterizing techniques for ablating tissue. Alternating current (AC) applied at a frequency of 490 kHz, known as radiofrequency (RF) energy, passes through prostatic tissue to cause resistive heating within the targeted areas and induce controlled coagulative necrosis at each treatment site.

A Comprehensive Guide to the Prostate. https://doi.org/10.1016/B978-0-12-811464-3.00017-9

Delivery of Radiofrequency Energy

TUNA, delivers RF energy into the surrounding prostatic tissue, causing intracellular heating and eventual necrosis. The tissue's resistance to the applied RF energy (modified AC current) results in hyperthermic cytotoxicity. RF is employed in a variety of other electrosurgical procedures, such as for the treatment of Wolff–Parkinson–White (WPW) syndrome. Catheters are used to access and deliver RF energy to the accessory atrioventricular bundles responsible for WPW symptoms.[4] Compared to higher-frequency RF applications like the microwaves used in transurethral microwave therapy (TUMT), the low-level RF applied with TUNA's monopolar instrument at 490 kHz provides deeper penetration and more even heat distribution within the tissue.

Experimental Studies

Initial tests on dog and cadaver prostates determined that 1 cm intraprostatic necrotic lesions could be achieved without damaging adjacent structures such as the rectum, bladder base, or urothelium.[5] Schulman et al. observed maximal necrosis of the targeted tissues occurring at 7 days after the procedure, with fibrosis developing at 15 days.[6]

Schulman and Zlotta demonstrated safety and efficacy of TUNA for the purpose of relieving symptomatic BPH.[7] Subsequent histochemical studies by Zlotta et al. on TUNA-treated prostatic tissue revealed a range of maximal lesion sized, from 10 mm × 7 mm to 20 mm × 10 mm. The ablations were localized to 0.3–1.0 cm from the urethra. This study also showed that tissue treated with TUNA did not stain for prostate-specific antigen, smooth muscle actin, or alpha adrenergic tissue markers.[8]

Instruments

The TUNA system (Medtronic, Inc, Minneapolis, MN) is comprised of a generator and hand-operated component. Its most updated iteration is called the Prostiva RF. While the earliest versions of the RF generators had to be manually adjusted by the urologist to optimize tissue heating, latest-generation TUNA systems monitor tissue impedance and temperature, automatically adjusting wattage to prevent the use of excessive energy.

Procedure

Patients are first given oral or IV sedation and positioned in the dorsal lithotomy position. 2% intraurethral lidocaine is then injected into the meatus, or a transperineal prostatic block is administered under ultrasound guidance. The dispersive electrode (also known as the grounding pad) is placed on the sacrum to complete the circuit. The most modern TUNA instruments are outfitted with a zero-degree fiber optic telescope, which allows direct visualization of the surrounding tissue and needle placement. Transrectal ultrasound (TRUS) can also confirm prostate shape and size and aid with the determination of needle length. Because the necrotic lesion can extend

5–6 mm beyond the needle tip, the electrode needs to be kept at an appropriately short length to protect the prostatic capsule and nearby organs. Once RF frequency is ready to be applied, a Teflon shield advances over the proximal part of the needle to protect the urothelium. Two needles project from the catheter through the surrounding urothelium and into the prostate (10 and 8 o'clock positions for the right lobe, and 2 and 4 o'clock positions within the left lobe), resulting in well-circumscribed lesions 1–2 cm in diameter, which are repeated down the length of the prostate. The target lesion core temperature for optimal necrosis is 115°C at which point a lesion will be generated in about 2 min.

Clinical Results

A mounting number of reported series and studies are showing TUNA to be comparable to TURP short-term in efficacy and safety. Currently, a lack of multicenter randomized clinical trials on this procedure limits the amount of data available to truly compare TUNA against other surgical, minimally invasive, or medical therapies for BPH.

Bouza et al. published a systematic review and metaanalysis of TUNA studies in 2006, concluding that TUNA has lower morbidity and an overall more favorable safety profile compared to TURP.[9] While quality of life and Qmax scores saw significant improvements from baseline in patients who underwent TUNA, the metaanalyses showed that by comparison, TURP remained the more effective therapy in objective and subjective improvements. Additionally, the long-term efficacy of TUNA was complicated by a high percentage of participants having to undergo either other surgical or medical retreatment. Analysis of noncomparative and comparative (TUNA vs. TURP) studies had rates of retreatment 19.07% and 10%, respectively.

In a RCT by Bruskewitz et al., IPSS improved by 30% for the majority (55%–89%) of patients.[10] Using a random effects model, Bouza's study calculated the percent improvement in symptom scores over a 2-year period for four studies, reporting a 61% improvement.[9]

Similarly to Bouza et al., the earlier Boyle and colleagues' metaanalysis on TUNA trials found that the quality of evidence supporting the introduction of TUNA into clinical use is lacking, citing a dearth of large RCTs with adequate follow-up times and recruitment rates.[11] The most recent studies highlight TUNA's excellent ejaculatory function outcomes when compared to TURP. Cimentepe and colleagues found patients treated with TUNA suffered a 0% rate of retrograde ejaculation.[12] In 2016, Marra et al. performed a systematic review on studies to investigate the effects of different surgical treatments on ejaculatory function, which showed significantly higher rates of retrograde ejaculation and decreased ejaculatory volume in patients who had TURP compared to TUNA.[13]

The American Urological Association (AUA) published a set of guidelines for BPH treatment in 2010, which was revalidated in 2014.[14] Per the guidelines, TUNA is effective at partially relieving BPH symptoms. TUNA as well as TUMT are listed in the AUA guidelines as minimally invasive treatment options to treat BPH.

Pressure-Flow Studies

The effect of TUNA on BOO by a decrease in detrusor pressure (Pdet) that was studied Campo et al., who observed an average reduction of 20%.[15] Steele et al. also showed similar drops in Pdet 2 years after TUNA.[16] A more recent randomized study by Roehrborn et al. measured obstruction parameters for 59 patients in the TUNA treatment group, observing a mean Qmax increase of 4.7 mL/s ($P<.0001$) and a statistically significant Pdet at peak Qmax of 18% at 6 months postprocedure.[17] Minardi et al. observed increased Qmax of 16.7%.[18] In 2007, a consecutive case series with 71 participants reported a mean Pdet at Qmax reduction of 19% at 3 and 12 months postprocedure.[19]

Cost-Effectiveness

Naslund et al. compared the costs of medical management or TURP versus TUNA and found that over a 5-year period, the least expensive treatment was monotherapy Tamsulosin ($3485).[20] 5 years of Finasteride monotherapy ($4867) was equivalent to a treatment with TUNA ($4811). The most expensive medical route was combination therapy ($4515), which reached a break-even point with TUNA ($4572) 2 years and 7 months posttreatment. The study found TUNA costs less than combination medical therapy after a 5-year period, reaching a break-even point at 2 years and 7 months.

Bouza and colleagues found a shorter hospital stay with TUNA patients versus TURP patients.[9] Conversely, the high retreatment rates experienced by TUNA patients could potentially have negative effects on the cost of this procedure, especially when compared to the price of alpha blocker monotherapy.

Adverse Effects

The adverse effect reported by the most patients in a majority of studies was hematuria.[9] In a metaanalysis performed by Bouza et al. on the studies comparing TUNA and TURP reported that the overall risk of adverse effects was consistently lower for TUNA. Cornu and colleagues' more recent open-label multicenter TUNA trial also reported hematuria in 12% of cases.[21] Serious complications included recurrent retention and retention with hematuria, which required rehospitalization and occurred in 2/49 patients in this study. Rosario et al. also reported mild hematuria in all cases, with 14% of subjects developing a UTI after the procedure.[19]

Reoperation

Varying rates of reoperation after TUNA have been reported in the literature. Rosario et al. found 83% of men reported a return of symptoms and required further intervention by the 20 month mark, although 17% of participants had no further intervention 10 years after undergoing TUNA.[19] This study concluded that the high retreatment rates for TUNA patients made it an expensive option for the treatment of BPH. The EAU Real-Life Data Registry established in 2003 reported a mean 5-year retreatment rate of 17.5%.[22] Bouza et al. observed that across the 17 studies reporting on additional interventions needed in the TUNA arm, there was 19.07% retreatment rate.[9]

Indications

Patients who would benefit the most from TUNA are those whose prostate volume is less than 60 g, with lateral lobe enlargement. While there exist reports of TUNA being performed on larger prostates and prostates with median lobe enlargement, data remain limited on its efficacy in these cases.[9] Unlike previous versions of TUNA, the needles in the Prostiva RF can be configured to reach enlargements of the median lobe, and treatment can be administered to this area so long as the needles do not advance past prostatic tissue into the bladder.

Summary

TUNA is a minimally invasive procedure, which achieves symptom relief comparable to other prostate ablation and resection methods, and which can mitigate the burdens of being on life-long medication. The flexibility offered by this in-office procedure also provides a treatment option for patients who would otherwise not be good candidates for TURP due to either surgical and anesthetic concerns or a desire to minimize the risk of retrograde ejaculation. Adverse outcomes are fairly uncommon. AEs such as hematuria, UTI, and AUR happen mainly because of improper needle positioning and difficulties with visualizing tissue depth. The risk of these complications is minimized with the use of TRUS and direct visualization. Most patients undergoing TUNA experience symptomatic relief and quantitative improvements in their Qmax and other indicators of bladder outlet flow, even if their initial presenting symptoms are severe. About 19% of patients may need additional medical or other surgical therapy within 2 years, which may affect treatment discussion when compared to other available treatments for BPH.

References

1. Jacobsen SJ, Girman CJ, Guess HA, Rhodes T, Oesterling JE, Lieber MM. Natural history of prostatism: longitudinal changes in voiding symptoms in community dwelling men. *J Urol* February 1996;**155**(2):595–600.
2. Lepor H. Landmark studies impacting the medical management of benign prostatic hyperplasia. *Rev Urol* 2003;**5**(Suppl. 5):S28–35.
3. Kaplan SA. Factors in predicting failure with medical therapy for BPH. *Rev Urol* 2005;**7**(Suppl. 7):S34–9.

4. Calkins H, Sousa J, el-Atassi R, et al. Diagnosis and cure of the Wolff-Parkinson-White syndrome or paroxysmal supraventricular tachycardias during a single electrophysiologic test. *N Engl J Med* June 06, 1991;**324**(23):1612–8.
5. Goldwasser B, Ramon J, Engelberg S, et al. Transurethral needle ablation (TUNA) of the prostate using low-level radiofrequency energy: an animal experimental study. *Eur Urol* 1993;**24**(3):400–5.
6. Vandenbossche M, Peltier A, Schulman CC. (TURF) transurethral radiofrequency heating for benign prostatic hyperplasia at various temperatures with Thermex II: clinical experience. *Eur Urol* 1993;**23**(2):302–6.
7. Schulman C, Zlotta A. Transurethral needle ablation of the prostate (TUNA): pathological, radiological and clinical study of a new office procedure for treatment of benign prostatic hyperplasia using low-level radiofrequency energy. *Arch Espanoles Urol* November 1994;**47**(9):895–901.
8. Zlotta AR, Raviv G, Peny MO, Noel JC, Haot J, Schulman CC. Possible mechanisms of action of transurethral needle ablation of the prostate on benign prostatic hyperplasia symptoms: a neurohistochemical study. *J Urol* March 1997;**157**(3):894–9.
9. Bouza C, Lopez T, Magro A, Navalpotro L, Amate JM. Systematic review and meta-analysis of transurethral needle ablation in symptomatic benign prostatic hyperplasia. *BMC Urol* June 21, 2006;**6**:14.
10. Bruskewitz RC. Quality of life and sexual function in patients with benign prostatic hyperplasia. *Rev Urol* Spring 2006;**5**(2):72–80.
11. Boyle P, Robertson C, Vaughan ED, Fitzpatrick JM. A meta-analysis of trials of transurethral needle ablation for treating symptomatic benign prostatic hyperplasia. *BJU Int* July 2004;**94**(1):83–8.
12. Cimentepe E, Unsal A, Saglam R. Randomized clinical trial comparing transurethral needle ablation with transurethral resection of the prostate for the treatment of benign prostatic hyperplasia: results at 18 months. *J Endourol* March 2003;**17**(2):103–7.
13. Marra G, Sturch P, Oderda M, Tabatabaei S, Muir G, Gontero P. Systematic review of lower urinary tract symptoms/benign prostatic hyperplasia surgical treatments on men's ejaculatory function: time for a bespoke approach? *Int J Urol* January 2015;**23**(1):22–35.
14. McVary KT, Roehrborn CG, Avins AL, et al. Update on AUA guideline on the management of benign prostatic hyperplasia. *J Urol* May 2011;**185**(5):1793–803.
15. Campo B, Bergamaschi F, Corrada P, Ordesi G. Transurethral needle ablation (TUNA) of the prostate: a clinical and urodynamic evaluation. *Urology* June 1997;**49**(6):847–50.
16. Steele GS, Sleep DJ. Transurethral needle ablation of the prostate: a urodynamic based study with 2-year followup. *J Urol* November 1997;**158**(5):1834–8.
17. Roehrborn CG, Burkhard FC, Bruskewitz RC, et al. The effects of transurethral needle ablation and resection of the prostate on pressure flow urodynamic parameters: analysis of the United States randomized study. *J Urol* July 1999;**162**(1):92–7.
18. Minardi D, Galosi AB, Yehia M, et al. Transurethral resection versus minimally invasive treatments of benign prostatic hyperplasia: results of treatments. Our experience. *Arch Ital Urol Androl Organo Uff Soc Ital Ecogr Urol Nefrol* March 2004;**76**(1):11–8.
19. Rosario DJ, Phillips JT, Chapple CR. Durability and cost-effectiveness of transurethral needle ablation of the prostate as an alternative to transurethral resection of the prostate when alpha-adrenergic antagonist therapy fails. *J Urol* March 2007;**177**(3):1047–51. discussion 1051.
20. Naslund MJ, Carlson AM, Williams MJ. A cost comparison of medical management and transurethral needle ablation for treatment of benign prostatic hyperplasia during a 5-year period. *J Urol* June 2005;**173**(6):2090–3. discussion 2093.

21. Cornu JN, Desgrandchamps F, de la Taille A, Vicaut E, Aout M, Lukacs B. Prospective short-term evaluation of transurethral needle ablation procedure in an ambulatory setting. *Urol Int* 2012;**89**(4):451–6.
22. Chappie AT, Hoefner K, Villavicencio H, Rosette JJDL. C.R. 887 Eau real-life data registry: long-term efficacy and safety of transurethral needle ablation (tuna@) in over 500 men with luts/bph. *Eur Urol Suppl* April 01, 2010;**9**(2):280.

Prostatic Urethral Lift

18

Karthik R. Krishnan, Dominique Thomas, Alexis Te, Bilal Chughtai
Weill Cornell Medicine, New York, NY, United States

The Urolift, developed by Neotract, is a minimally invasive device that alleviates benign prostatic hyperplasia (BPH)–induced lower urinary tract symptoms (LUTS). Approved for LUTS treatment by the FDA in September 2013, it is also available in the European Union, Australia, New Zealand, Canada, Serbia, and Turkey. The name "Urolift" describes both the device and procedure, known as the prostatic urethral lift (PUL).

In contrast to most other minimally invasive surgeries for BPH/LUTS, the Urolift operates mechanically instead of relying on prostate tissue removal. Furthermore, it does not conform to the shape of the prostatic urethra (PU) such as prostatic urethral stents and other mechanical solutions.

During a PUL procedure, the Urolift prostheses are endoscopically guided and deployed to direct the PU toward the prostate capsule to widen the prostatic urethral space. This is accomplished by connecting several pairs of metal tabs with one set deployed inside of the PU and the other on the capsule exterior. Nonabsorbable sutures connect each pair of tabs, opening up the PU under tension. After deployment, the Urolift is subsequently enveloped by the urothelium and prostate tissue.

The Urolift System

The Urolift urethral and capsular tabs are composed of stainless steel and nitinol, respectively. The urethral tab is positioned along a size 0 polyethylene terephthalate (PET) nonabsorbable monofilament suture during the procedure. The implant system then cuts the suture to the required length.

A single Urolift implant is loaded into each delivery system, meaning multiple systems must be used to deliver multiple implants. A 20F cystoscope sheath, is used to guide the delivery system. Visualization during the PUL is achieved through a 0 degree 2.9 mm cystoscope lens. A surgeon may handle and manipulate the system as they would a resectoscope. The sequence of events for successful deployment are as follows:

1. With the first trigger pull, the needle and a capsular tab loaded at the needle's end are deployed through the prostate to outside the capsule.
2. Completely depressing the trigger retracts the needle. The capsular tab is released and secured outside the prostatic capsule as the suture comes under tension.
3. Pushing the thumb-operated button along the outside edge of the handle further tenses the suture, sets the urethral tab in place on the suture, and cuts the suture flush with the urethral wall.

A Comprehensive Guide to the Prostate. https://doi.org/10.1016/B978-0-12-811464-3.00018-0

Prostatic Urethral Lift Procedure

Several types of anesthesia may be used to perform the Urolift, including topical lidocaine, prostate blocks, or general anesthesia depending on patient and surgeon preferences. The PUL is best performed with the patient in the dorsal lithotomy position. The following steps:

1. Rotate the handle of the device as necessary to position the tabs along the anterolateral borders (2 and 10 o'clock positions) of the PU.[1] This allows the surgeon to create an anterior channel in the PU while avoiding damage to the neurovascular bundle or dorsal venous complex.[1]
2. After finding a suitable deployment location, finer rotational adjustments may be made while maintaining contact between the device tip and tissue.
3. Remove the safety lock and pull the trigger.
4. Completely depress the trigger. The suture will now be visible.
5. Straighten the device and align it parallel to the urethra.
6. Advance the device forward slowly until the suture reflects light, and then depress the thumb button.
7. Evaluate whether the lateral lobes have been retracted suitably. Repeat steps 1–6 until the anterior channel becomes visibly clear.

Clinical Studies

Initial Urolift data were collected at two Australian institutions using two prototypes of the current commercial iteration.[3] In a study by Woo et al., a total of 19 patients were assessed. This study demonstrated that the device could deploy the implants across the PU and the prostate capsule. These procedures were extensively monitored through a combination of visual endoscopic, fluoroscopic, and transrectal ultrasound guidance. After the feasibility of transprostatic implant delivery was established, the following PULs were performed without fluoroscopy or ultrasound. These feasibility studies demonstrated the safety of this procedure.

The study was expanded to include six Australian sites with 64 patients over a 2-year follow-up.[4] Three separate prototypes of the current system were utilized for the first 25 cases while the procedure was being refined. Subsequent cases used the same delivery system as cases 20–25.

At 2 weeks, International Prostate Symptom Scores (IPSS) dropped from an initial mean of 22.6 ± 5.4 to 13.2 ± 6.3 (42% reduction). This reduction persisted through the 2-year follow-up end point (12.6 ± 7.2). Smaller improvements were seen in peak flow rate. Paired analysis revealed improvements from 8.3 ± 2.2 mL/s (baseline) to 12.0 ± 7.6 mL/s (2 weeks) and from 7.4 ± 2.2 mL/s (baseline) to 10.3 ± 4.1 mL/s (2 years).[4] Similar improvements were seen in the Quality of Life (QoL) index and the

[1] The surgeon should avoid applying too much pressure to the prostate to avoid impacting the pubic bone.

BPH Impact index both at 2 weeks and 2 years out. The postresidual volume did not demonstrate any statistically significant changes.

In a 2-year follow-up study, 20% of men required retreatment (13/64). Out of these patients, 12 received photoselective prostate vaporization or transurethral prostate resection. The remaining patient opted for a second Urolift procedure. Within this group, 10 patients were from the first 25 patients (using older techniques) while only 3 came from patients 26–64 (using the current methods).

The "Luminal Improvement Following Prostatic Tissue Approximation for the Treatment of LUTS secondary to BPH" (LIFT) study was the first randomized control trail (RCT) to compare PUL against a sham procedure.[5] LIFT was a randomized, international blinded study across 19 centers. Initially 206 men were randomly assigned to either PUL or sham study groups in a 2:1 ratio.

After 3 months, subjects were unblinded to the procedure they received and were given the option to receive active treatment if they initially received a sham procedure. The LIFT end point was demonstration of a significant (25% greater) improvement in American Urological Association Symptom Index (AUA-SI) score relative to the sham group with the intention to treat. The PUL group showed improvement in AUA-SI scores from 22.2 ± 5.48 to 11.2 ± 7.65, while the sham group only experienced a score drop from 24.4 ± 5.75 to 18.4 ± 8.59.[5] The results at 1 year revealed similar improvements in symptoms.

Of the 66 men in the sham group, 53 opted to receive PUL treatment at 3 months. The paired comparison between sham and PUL data from these men at 3 months demonstrated similar improvements.

Follow-up LIFT studies were conducted to assess the durability of these changes at 2 and 3 years after the initial procedure.[4,6] All measured improvements in IPSS, urinary flow, and quality of life remained stable through the follow-up. In addition, there was no decrease in efficacy with an increase in prostatic volume.

The BPH6 study was a RCT comparing, TURP, with PUL in 91 men. Participants were evaluated at 2 weeks, 1, 3, 6, and 12 months.[7] This study measured (urinary continence, sexual function), intraoperative and postoperative safety, clinical efficacy,[7] BPH Impact Index, peak flow rate, Male Sexual Health Questionnaire for Ejaculatory Dysfunction (MSHQ-EjD), and the Sexual Health Inventory for Men (SHIM).

The study's primary Endpoint was BPH6: a composite of six secondary clinical endpoints. This score was defined by the percentage of participants in each study group that met all six secondary criteria.

At 12 months, PUL outperformed TURP in the following: postoperation recovery (82% vs. 53%) and Ejaculatory Function (100% vs. 61%). Analyzed together, 53% of PUL patients met the BPH6 Primary Endpoint compared to 20% of TURP patients.[7] With regards to other parameters, both procedures resulted in significant, comparable improvements in Qmax, IPSS, IPSS QoL, and BPH Impact Index. TURP outperformed PUL in improving Qmax while PUL resulted in better SHIM and MSHQ-EjD scores.

A 2-year follow-up study found that these changes persisted through 24 months across both the PUL and TURP treatment groups.[8] No statistically significant difference was seen between the two study groups when comparing improvements/maintenance of Health-Related Quality of Life, postvoid residual urine volume, SHIM, or

MSHQ-EjD Bother. TURP continued to show larger improvements in IPSS and Qmax while PUL better preserved ejaculatory function (MSHQ-EjD function). Patient also had a statistically significant improvement in sleep quality in the PUL group based on the Jenkins Sleep Questionnaire.

Further studies have corroborated these findings in a study involving four men treated with PUL demonstrated a median procedure time of 11 min (range 6–11 min).[9] One patient showed no urinary function improvement while the others experienced improvements similar to those in prior studies. The same benefits were seen in a 20-patient study larger prostates up to 109 cc.[10] Another trial involving 51 men, which mimicked the data instruments and inclusion criteria of the LIFT study, found similar improvements and results when analyzing symptoms, flow rate, and sexual function.[11]

Indications

Usage of the Urolift is indicated for men with LUTS secondary to BPH. It is not contraindicated in patients who experienced past failure of pharmacotherapy or PULs. Prostate anatomy could potentially affect the efficacy of the Urolift given that most of the clinical trials described did not include glands with large middle lobes or those greater than 80 cc in volume. As such, further studies are required to establish Urolift's efficacy or lack thereof in such patients. Although the safety of the Urolift was not discussed in these trials, it was tested in standard clinical use conditions. In addition, it is designated as "MR conditional" with respect to MRI compatibility.

Adverse Events

Clinical data compiled so far suggest a lower morbidity associated with PUL relative to other surgical LUTS/BPH techniques. Adverse events, including bladder spasms, dysuria, and hematuria, were temporary, lasting up to 2–4 weeks.[12] No serious adverse clinical events have been reported in the literature.

Sexual Function

Clinical trials testing the Urolift have not reported any cases of sexual dysfunction postprocedure after evaluation with clinically validated tools.[6] In comparison, the vast majority of alternative treatments for BPH/LUTS, including other MISTs, pharmacotherapy, and cavitating surgery, are associated with measurable levels of ejaculatory dysfunction.[10]

The multicenter Australian study measured changes in sexual dysfunction using MSHQ-EjD and SHIM at each evaluation time point (baseline, 6 weeks, 3, 6, and 12 months). The results showed no decrease in function out through 1 year. These metrics were used in the LIFT study as well, where it was discovered that the men overall experienced consistent improvements in ejaculatory function. These results were replicated in the BPH6 study with the same metrics.

Future Directions

Further studies examining Urolift are needed to understand longer-term outcomes. Further, the role of those with median lobes will help expand its effectiveness.

References

1. McNicholas TA, Woo HH, Chin PT, et al. Minimally invasive prostatic urethral lift: surgical technique and multinational experience. *Eur Urol* August 2013;**64**(2):292–9.
2. Barkin J, Giddens J, Incze P, Casey R, Richardson S, Gange S. UroLift system for relief of prostate obstruction under local anesthesia. *Can J Urol* April 2012;**19**(2):6217–22.
3. Woo HH, Chin PT, McNicholas TA, et al. Safety and feasibility of the prostatic urethral lift: a novel, minimally invasive treatment for lower urinary tract symptoms (LUTS) secondary to benign prostatic hyperplasia (BPH). *BJU Int* July 2011;**108**(1):82–8.
4. Chin PT, Bolton DM, Jack G, et al. Prostatic urethral lift: two-year results after treatment for lower urinary tract symptoms secondary to benign prostatic hyperplasia. *Urology* January 2012;**79**(1):5–11.
5. Roehrborn CG, Gange SN, Shore ND, et al. The prostatic urethral lift for the treatment of lower urinary tract symptoms associated with prostate enlargement due to benign prostatic hyperplasia: the L.I.F.T. study. *J Urol* December 2013;**190**(6):2161–7.
6. Roehrborn CG, Rukstalis DB, Barkin J, et al. Three year results of the prostatic urethral L.I.F.T. study. *Can J Urol* June 2015;**22**(3):7772–82.
7. Sonksen J, Barber NJ, Speakman MJ, et al. Prospective, randomized, multinational study of prostatic urethral lift versus transurethral resection of the prostate: 12-month results from the BPH6 study. *Eur Urol* October 2015;**68**(4):643–52.
8. Gratzke C, Barber N, Speakman MJ, et al. Prostatic urethral lift vs transurethral resection of the prostate: 2-year results of the BPH6 prospective, multicentre, randomized study. *BJU Int* May 2016;**119**(5):767–75.
9. Delongchamps NB, Conquy S, Defontaines J, Zerbib M, Peyromaure M. Intra-prostatic UroLift((R)) implants for benign prostatic hyperplasia: preliminary results of the four first cases performed in France. *Progres en urologie: Journal de l'Association francaise d'urologie et de la Societe francaise d'urologie* September 2012;**22**(10):590–7.
10. Garrido Abad P, Coloma Del Peso A, Sinues Ojas B, Fernandez Arjona M. Urolift(R), a new minimally invasive treatment for patients with low urinary tract symptoms secondary to BPH. Preliminary results. *Arch Espanoles Urol* July–August 2013;**66**(6):584–91.
11. Shore NA. Review of the prostatic urethral lift for lower urinary tract symptoms: symptom relief, flow improvement, and preservation of sexual function in men with benign prostatic hyperplasia. *Curr Bladder Dysfunct Rep* 2015;**10**:186–92.
12. *NeoTract, Inc. Announces data showing excellent durability of results for UroLift system at four years*. UroLift; 2016.

Prostatic Stents

19

Carrie M. Mlynarczyk[1], *Erica L. Ditkoff*[2], *Gina M. Badalato*[1], *Matthew P. Rutman*[1]
[1]New York Presbyterian Hospital/Columbia University Medical Center, New York, NY, United States; [2]Brigham & Womens Hospital, Boston, MA, United States

The Role for Prostatic Stents

Lower urinary tract symptoms caused by BPH are common in men and increase with advancing age. Approximately one-third of patients fail maximal medical therapy for bladder outlet obstruction (BOO) from BPH.[1] The traditional procedure of next resort for these patients is a transurethral resection of the prostate (TURP), which historically was associated with a high morbidity and mortality rate of 18% and 0.2%, respectively. Contemporary data have shown a significant improvement in outcomes, with a reported mortality of 0.1% during the first 30 days after the procedure. Other perioperative complications include TUR syndrome occurring in 0.8% of patients (range 0%–5%), blood transfusion in 2% (range 0%–9%), clot retention in 4.9% (range 0%–39%), and urinary tract infection in 4.1% (range 0%–22%).[2–4]

While these complication rates are low, a subset of men with significant comorbidities have an increased risk for worse outcomes. In men over 80 years of age, the age-adjusted morbidity and mortality rates after TURP are estimated at upward of 25% and 0.2%, respectively. In addition, 10%–15% of all elderly men are determined to be poor candidates for a TURP.[4,5] With increasing life expectancy and the aging baby boomer population these issues have growing importance.[1] The Center for Disease Control and Prevention report that there is an unprecedented increase in the proportion of older adults in the population. The US census issued a report detailing that the population over 65 years of age will double between 2012 and 2050.[6]

Less invasive interventional therapies provide the benefit of lower morbidity and mortality. High-risk patients may be unable to undergo a definitive surgical procedure due to chronic obstructive pulmonary disease, congestive heart failure, ischemic heart disease, or a need for prolonged anticoagulation therapy. Transurethral vaporization of the prostate, holmium laser resection, and laser ablation of the prostate require general or spinal anesthesia.[2] Some treatments can be performed under local anesthesia or sedation in an attempt to modulate anesthetic risk, and these include UroLift, Rezum water vapotherapy, transurethral microwave thermotherapy (TUMT), transurethral needle ablation, and interstitial laser coagulation. However, the advantages of these office-based procedures must be tempered with outcomes that include a lower postoperative flow rate and higher incidence of reoperation.[7–9] The UroLift system is a relatively new minimally invasive technique that can be placed under local anesthesia or with sedation; however, it cannot be used in men with relatively large prostates (>80 g) or a prominent median lobe component. Furthermore, long-term durability data for

A Comprehensive Guide to the Prostate. https://doi.org/10.1016/B978-0-12-811464-3.00019-2

this procedure beyond 4 years are lacking.[10] Comparatively, placing a prostatic stent has no exclusion criteria based on prostate morphology, and it is a relatively short outpatient procedure that can be done under local or monitor-assisted anesthesia; as will be discussed in subsequent units, patients sustain dramatic improvements in quality of life and voiding parameters in the immediate postoperative period.

Furthermore, prostatic stents can be an alternative to CIC or an indwelling catheter in patients who are not surgical candidates. There are known disadvantages and risks to CIC and Foley catheter placement. CIC requires a certain level of patient functional and cognitive capacity and has an inherent risk of traumatic catheterization in the setting of BPH. An indwelling catheter, suprapubic or urethral, is also associated with intrinsic problems, including the negative impact on patient quality of life associated with an external drainage device and the higher independent risk of infection.

History

The use of stents to relieve obstruction is used in many medical fields. Vascular surgeons initially developed stents in the early 20th century.[11] They were introduced for use in cardiac and vascular diseases. In 1969 Dotter detailed early work in end-arterial tube grafts with the goal of improving outcomes through minimally invasive techniques. His work stressed the use of stents to forego more traumatic, open procedures.[11] The use of stents has expanded from vascular surgery and has been used in otolaryngology, ophthalmology and gynecology. For example, in 1989 Hurwitz detailed a promising experience with Teflon tear duct tubes.[12] Furthermore, in 1999, Mair et al. reported on the use of endobronchial stents for the treatment of severe bronchomalacia in patients unresponsive to maximal medical management.[13] Authors across multiple specialties thus comment on the utility of stent insertion as a minimally invasive intervention that can prevent the need for more complex surgical procedures.

Fabian was the first to describe a stent adapted as therapy for prostate obstruction. In 1980 he published on his development of the "urological spiral" to temporarily relieve BOO.[14] When urological stents were initially introduced, the treatment was applied to the general BOO population and the variable outcomes and side effects deterred its ongoing application. Nevertheless, although this chapter will focus on the use of prostatic stents for BPH, contemporary urethral stents have applications for other forms of obstructive uropathy such as urethral stricture disease and detrusor sphincter dyssynergia.

A prostatic stent is a tube that can be endoscopically placed into the prostatic urethra to tent open prostatic tissue and to relieve BOO, allowing for normal voiding in patients with a functioning detrusor.[15]

There are two classifications of stents: temporary and permanent. Permanent stents allow for tissue ingrowth and the stent becomes anchored into the urethral wall similar to how endovascular stents ultimately anchor to vessel walls. Temporary stents are composed of either nonabsorbable or biodegradable material. Unlike permanent stents, temporary stents do not become epithelialized within the urethra. Ideally, this characteristic prevents difficulty in removing temporary stents and allows the removal procedure to be performed under topical anesthesia with sedation, rather than under general anesthesia. The duration that a temporary stent can remain in situ varies based on the stent type.[16]

The goals of the prostatic stent placement procedure are to relieve BOO and to improve quality of life without causing incontinence or requiring general anesthesia or a complicated insertion technique.[16] There are limited data on the efficacy and safety of prostatic stents. Most studies originate from the European literature. Early studies showed promise; however, patients experienced multiple complications, and the products continue to undergo various revisions with multiple generations of stent development. Permanent prostatic stents have been removed from the North American market and the Spanner Temporary Prostatic Stent (SRS Medical, N. Billerica, MA, USA) is the stent that is currently FDA-approved.

Permanent Stents

First Generation

The UroLume (American Medical Systems, Minnetonka, Minnesota, USA): The UroLume Wallstent introduced in 1988 was one of the first permanent stents. It is composed of stainless steel alloy woven into a 42 French lumen tubular mesh.[17] This stent was studied in men deemed unfit for surgery due to advanced age or comorbidities.[16] While initial trials were promising, further follow-up suggested a high rate of explantation. An early multicenter prospective study of 126 men who received the UroLume found improved voiding function with an overall removal rate of 13% at 2 year follow-up.[18] The most comprehensive review of the UroLume, a 2007 metaanalysis of 990 patients with BPH, suggested good stent efficacy. Eighty-four percent of subjects who were catheter dependent were able to void spontaneously after placement with an increase in their average flow rate from 4.2 to 13.1 cc/s. Of the 606 subjects who had 1 year of follow-up, 104 (16%) sustained stent failure with stent migration as the largest cause of failure (38 cases, 37%).[19] However, a long-term follow-up of 62 patients treated with the UroLume highlighted the high rate of explantation. In this study 22 and 11 patients completed 5 and 12 year follow-up, respectively. Of the 29 (47%) stents explanted, 18 were removed within the first 2 years due to either migration or inappropriate placement; and a total of four stents were removed due to lack of efficacy. Of the stents removed, 72% were removed in the first 2 years. Additionally late stent removal was seen in patients who developed recurrent outlet obstruction from encrustation of the stent. This study did find that 34% of men died of unrelated causes with the stent in place.[5] This suggests that with care in patient selection, there is a subgroup that could benefit from the UroLume. Critics suggest that a significant problem with the stent was that the round shape and 42 French (just to be consistent with previous writing) size did not adequately fill all portions of the lumen of the urethra.

Second Generation

The second generation of stents was designed in response to the significant problems caused by stent migration.

Memokath (Engineers and Doctors A/s, Hornbaek, Denmark): Described in 1993, it was made of titanium nickel alloy to allow for improved urethral fit through temperature control. When hot water was passed through the stent a portion would expand into the prostatic apex to maintain the correct placement, and when flushed with cold water the stent would soften and allow for removal.[17] This stent could be placed with a catheter delivery system under ultrasound guidance. In an initial study of 30 patients with BPH they found an 83% success rate, however, their study had limited follow-up of 0.2–9 months with a median of 3 months.[20] A smaller series of 15 patients with a mean age of 87 years and with American Society of Anesthesiologists grade of 3 or 4 showed 9 (60%) catheter-free at 1 year, 3 deceased from unrelated causes, and 3 with stent removal at a mean of 9 months.[21] A metaanalysis of 839 patients found that the Memokath stent reduced International Prostate Symptom Scores (IPSS) by 11–19 points and resulted in a Qmax increase of 3–11 mL/s. However, this study assessed symptoms at variable time points after placement and follow-up was insufficient to comment on durability.[22] Another study followed 211 men who were high-risk for surgery or refused surgery. The stent was placed under local anesthesia. Over 8 year follow-up 38% of the patients died with their stent in place and 34% remained alive with a functioning stent. 23% had the stents removed, 4% of which were done because the patients were no longer unfit for surgery.[23]

Memotherm (Angiomed, Germany): This stent is similar to Memokath with a nitinol wire mesh thermosensitive design. First described in 1993, this stent design was flexible and gained maximal expansion at body temperature with the goal of adapting to the prostatic urethra. The knitted structure was designed to allow for removal. Multiple studies have been done with varying results. A study from Germany of 54 patients with a mean age of 76 years found an improvement in maximum flow, postvoid residual, and American Urological Association symptom score.[24] A study of 15 patients out of Tokyo saw improvement in 12 of the patients' urinary function. However, two patients were unable to urinate, and one stent was explanted for stone formation.[25] A study from London followed 48 men with Memotherm placement for 6 months. 47 patients were able to void; however, only 15 had satisfactory results, 10 stents were explanted, with 3 additional explants planned.[26] Gesenberg et al. assessed this stent with the goal of obtaining long-term follow-up. Due to significant comorbidities many of the patients were unable to return for the full duration of follow-up, and 36% of the patients died from unrelated causes. The average follow-up time was 12.2 months. In the group that had 4-year data, they did see an improvement in average urinary peak flow. Early stent dislocation was discovered in 9% and corrected with a repositioning tool. Five stents were removed, 19 stents were obstructed due to further hyperplasia requiring additional interventions, 13 patients had urethral strictures requiring procedures, and 4 patients developed stone formation on the stents.[27] While most of the studies lack sufficient long-term information, Gesenberg et al. highlight the difficulty of obtaining follow-up in this patient population due to their significant comorbidities and limited life expectancy.

Hourglass/Bell-shaped nitinol urethral stent: After continued problems with migration in the temperature controlled stents, van Dijk et al. published on their experience with altering the shape of the stent. Additionally, this stent was placed under cystoscopic guidance to ensure correct positioning. In a study of 35 men they placed a thermal hourglass-shaped stent, which had an increased diameter at both ends, with the aim of anchoring the stent in place. However, despite the size modification, they continued to see a high rate of migration requiring removal: at 12 months only 23% of the stents were in place. Despite the high migration rate, the patients saw a significant initial improvement in voiding and symptom scores, highlighting the importance of sufficient follow-up before claiming success of a stent.[28] The author's hypothesized that the problem with the configuration was that the larger curved end, toward the bladder, facilitated the migration. A second study with a bell-shaped stent included 108 men. Again they found initial improvement in symptom score and voiding parameters yet continued to see a high rate of stent explantation. The median time to stent removal was 105 days and by 30 months only 18% of the stents remained in place.[29]

Permanent Prostatic Stent Removal

A range of complications including stent migration, return of symptoms, intolerability, or encrustation necessitated a consistently high rate of removal. The North American Study Group reported on their experience of removal of 73 stents from 69 patients. They found that 43.8% of the explantations were done in the first year. For stent removal in the setting of BPH, they detailed using low-cutting, monopolar current with a standard resectoscope. They used one of two techniques for removal: (1) alligator forceps were used to stretch and narrow the stent to allow for removal by the resectoscope; or, (2) a wire was placed through the stent, the device pushed into the bladder, and then removal completed by pulling on the ends of the wire.[30]

A subsequent study detailed the removal algorithm for the UroLume stent. The initial attempts followed the manufacturer's instructions to remove overlying urothelium with a resectoscope, push the stent into the bladder, and remove it en todo through a larger sheath. If that was not successful, then the authors used a grasper to attempt to remove the exposed stent. The final method involved removing the composite wires individually under fluoroscopic guidance.[31] The above procedures were complicated and likely would require general anesthesia thus potentially defeating the purpose of stent use in the first place.

The Memokath stent strove to address this problem with the thermosensitive design. The stent was made with the capability of cold saline to soften and uncoil the stent thus facilitating removal. A multicenter study out of Europe reported on their experience of Memokath removal. Of the 93 patients undergoing stent explantation, 38% received general anesthesia, 46% received topical anesthesia, and 14% received regional anesthesia. The mean duration of the procedure was 11 min. While the average time of the Memokath removal was shorter than a standard TURP or laser procedure, over a third required general anesthesia.[32]

Temporary Stents

Temporary stents were developed primarily for patients with transient BOO. Transient BOO may occur after recent prostate surgery, including brachytherapy or high-energy TUMT, as these procedures cause temporary tissue edema. For management of the resultant urinary obstruction, a prostatic stent is an alternative to indwelling Foley catheter, suprapubic tube, or self-catheterization.[17]

First Generation

These spiral stents require removal 12 months after their placement into the prostatic urethra. They are placed into the prostatic urethra either with endoscopic or ultrasound guidance and the use of a catheter guide.

Prostakath (Pharma-Plast): In 1993 Ala-Opas and colleagues reported on a multicenter study that assessed the usefulness of the Prostakath for managing urinary outflow obstruction. Eighty-seven males were included in their study. Of these patients, 68 had complete urinary retention. 81% (69) had clearance of their outflow obstruction. However, during follow-up 39% (33) of the stents were removed. Of the stents removed, 15 were removed within a few days of placement. The mean functioning time of the Prostakath was 9 months.[33] The same year that Ala-Opas et al. published these results, Thomas and colleagues reported findings from a similar, but smaller, study. Sixty-four patients in their cohort presented with acute urinary retention and had Prostakath stents placed. Following the procedure, 89.1% (57) were able to void. A quarter of the patients (16) experienced stent-associated problems, with the most common complication being displacement. Of the patients with functioning stents, the mean follow-up period was 12.0+9.9 months.[34] While these studies concluded that the Prostakath can improve voiding they continued to see a high rate of removal and stent-associated complications.

Urospiral (Porges): The Urospiral is a first-generation temporary intraprostatic stent. It is composed of stainless steel and designed as three parts: several spiral loops in the prostatic urethra, a 2 cm straight wire in the membranous urethra, and a distal two loops in the bulbar urethra. Ozgur et al. followed 31 patients treated with Urospiral from May 1989 to September 1991. All were deemed to be high risk for surgery and 14 had indwelling catheters. Using local anesthetic the stent was introduced under cystoscopic guidance over a ureteral catheter and placed under fluoroscopic guidance. All patients reported subjective improvement and demonstrated complete bladder emptying. No patients reached normal flow rates. Two patients experienced minor incontinence requiring pads. Four patients had symptomatic urinary tract infections that were treated with antibiotics.[35]

Second Generation

In the late 1980s, second-generation intraprostatic stents replaced those from the first generation. The second-generation stents were designed to have larger internal diameters to reduce rates of spontaneous migration and stent occlusion by stones or clots.[36]

ProstaCoil: ProstaCoil is a second-generation temporary and self-expanding coil stent; it is composed of nickel–titanium alloy and has shape memory. The stent can be inserted at a caliber of 17 French and then expanded to a maximum caliber of 24/30 French.[37] In 1996 Yachia and Aridogan compared the ProstaCoil to the ProstaKath. The study included 117 patients ranging 52–94 years of age with prostatic obstruction.

86 patients received the ProstaCoil, while 49 patients received the ProstaKath. In 10% of the patients who were treated with the ProstaKath, the stent could not be inserted due to stent instability. The ProstaCoil was correctly placed on first attempt 100% of the time and no migration was observed. The ProstaKath was correctly positioned immediately 83% of the time, 42% of the patients required later stent repositioning secondary to partial migration, and 12% required stent removal because of complete migration. It is likely that the reduced occurrence of ProstaCoil migration was secondary to the larger caliber. However, the ProstaCoil did result in more transient irritative symptoms. At 1-year follow-up, 40% of the ProstaKath group had stent encrustation, while 30% of the ProstaCoil group had stent encrustation at 2 years. There was no difference in stent-induced infections across the two groups. Maximal stent indwelling time was 1 year for the ProstaKath and 3 years for the ProstaCoil. Overall, the ProstaCoil was superior to the ProstaKath in terms of ease of placement, migration and encrustation rates, and maximal duration of use for treating patients with prostatic obstruction.[36]

The Spanner (Abbey Moor Medical, Inc., Minnesota, and USA): The Spanner is a retrievable stent that essentially constitutes the proximal 4–6 cm of a Foley catheter that encompasses the length of the prostatic urethra. The device can remain in place for a mean of 57 days, and pilot data involving 33 subjects originally reported a 68% decrease in IPSS (from 22.3 to 7.1) and 42% improvement in average flow rate (8.2–11.6 mL/s). Lack of device migration was confirmed radiographically at 12 weeks.[38]

There may be additional benefit in the postbrachytherapy and TUMT settings.[39] A trial of 196 subjects randomized patients to a stent (100) or standard of care, which included routine Foley catheterization (86) immediately following TUMT. The patients had a stent/Foley in place for up to 8 weeks; the Spanner group reported greater improvements voiding symptoms as measured by IPSS scores that were sustained following stent removal.[40]

Allium Triangular Prostatic Stent: The Allium Triangular Prostatic Stent (TPS) is a relatively new temporary stent. The stent is flexible and composed of nitinol and covered with copolymer to prevent ingrowth of tissue, encrustation, calcification, and stone formation. Unlike most other stents that are circular, the Allium TPS has a triangular structure. The triangular structure minimizes the amount of prostatic tissue that is displaced by the stent and thus reduces resultant irritative voiding symptoms. In 2016 Yildiz and colleagues reported the results of the Allium TPS use in patients with benign prostatic obstruction (BPO) who were not surgical candidates secondary to patient preference or comorbidities. In the study, 51 patients underwent Allium TPS placement between 2008 and 2014. Stent insertion was successful in all patients under local or regional anesthesia, and patients were followed for 12 months. Mean IPSS scores decreased from 26.4 to 7.7 over the follow-up period. Additionally, mean peak urinary flow increased to 16.0–5.5 mL/s. Patients reported experiencing improvement in quality of life following stent placement. No stent migration or obstruction was observed; however, 17.6% (9) of patients reported transient pain. Overall, the authors concluded that the Allium TPS is safe and effective in the short-term for patients with BPO.[41] More long-term follow-up is needed to adequately evaluate the risk of stent malfunction requiring explantation.

Biodegradable Stents: A biodegradable stent is the ideal temporary stent because it dissolves in situ and thus does not require removal.[42,43] Although this concept is appealing, the clinical efficacy of these prototypes has yet to be documented. A rabbit study out of Finland explored the suitability of biodegradable self-reinforced poly-L-lactide stents. The study's authors found that there was a decreased level of inflammatory reaction compared to steel.[42] A study of 45 patients evaluated the efficacy and safety of a biodegradable self-reinforced poly-DL-lactic acid spiral stent after visual laser ablation of the prostate. The patients all completed 6-month follow-up. They found earlier voiding in the stent and suprapubic tube group than in the suprapubic tube group. However, they found no significant differences between the groups for maximum urinary flow rate, postvoid residual, or symptom score. On cystoscopic exam, the stents were intact and in position at 1 month, partially degraded at 3 months, and fragmented in the bladder or urethra at 6 months. Two of the patients in the spiral stent group formed stones.[43]

Conclusions

Among elderly men who seek symptomatic improvement from severe BPH-related BOO and who are not candidates for a transurethral surgical procedure, permanent prostatic stents may have an acceptable short-term application. However, the difficulties and complications inherent to epithelializing stents, including encrustation, tissue ingrowth, obstruction, and migration, prohibit widespread long-term usage in this patient population. Retrievable prostatic stents, specifically the Spanner, have been shown to have temporizing applications following procedures that result in prostatic tissue edema. The Allium stent's early research is promising but requires longer follow-up. Biodegradable stents remain experimental.

As surgical procedures for transurethral BOO with low morbidity and mortality continue to develop, the additional benefit provided by prostatic stents is difficult to demonstrate. Nonetheless, with an aging population and increasing life expectancy there is a subset of patients who may benefit from prostatic stent placement, an office-based procedure that can be done under local anesthesia and relieve symptoms of BOO.

References

1. Roehrborn CG. Male lower urinary tract symptoms (LUTS) and benign prostatic hyperplasia (BPH). *Med Clin N Am* 2011;**95**(1):87–100.
2. Madersbacher S, Marberger M. Is transurethral resection of the prostate still justified? *BJU Int* 1999;**83**(3):227–37.
3. Ahyai SA, Gilling P, Kaplan SA, et al. Meta-analysis of functional outcomes and complications following transurethral procedures for lower urinary tract symptoms resulting from benign prostatic enlargement. *Eur Urol* 2010;**58**(3):384–97.
4. Mebust WK, Holtgrewe HL, Cockett AT, Peters PC. Transurethral prostatectomy: immediate and postoperative complications. A cooperative study of 13 participating institutions evaluating 3,885 patients. *J Urol* 1989;**141**(2):243–7.
5. Masood S, Djaladat H, Kouriefs C, Keen M, Palmer JH. The 12-year outcome analysis of an endourethral wallstent for treating benign prostatic hyperplasia. *BJU Int* 2004;**94**(9):1271–4.

6. Ortman JVV, Hogan H. An aging nation: the older population in the United States: population estimates and projections. In: Bureau USDoCEaSAUSC, editor. *Current population reports*. United States Census Bureau; 2014. https://www.census.gov/prod/2014pubs/p25-1140.pdf.
7. Hoffman RM, Monga M, Elliott SP, et al. Microwave thermotherapy for benign prostatic hyperplasia. *Cochrane Database Syst Rev* 2012;(9):Cd004135.
8. Bouza C, Lopez T, Magro A, Navalpotro L, Amate JM. Systematic review and meta-analysis of transurethral needle ablation in symptomatic benign prostatic hyperplasia. *BMC Urol* 2006;**6**:14.
9. Martenson AC, De La Rosette JJ. Interstitial laser coagulation in the treatment of benign prostatic hyperplasia using a diode laser system: results of an evolving technology. *Prostate Cancer Prostatic Dis* 1999;**2**(3):148–54.
10. Chin PT, Bolton DM, Jack G, et al. Prostatic urethral lift: two-year results after treatment for lower urinary tract symptoms secondary to benign prostatic hyperplasia. *Urology* 2012;**79**(1):5–11.
11. Dotter CT. Transluminally-placed coilspring endarterial tube grafts. Long-term patency in canine popliteal artery. *Investig Radiol* 1969;**4**(5):329–32.
12. Hurwitz JJ. Teflon tubes for stenting and bypassing the lacrimal drainage pathways. *Ophthalmic Surg* 1989;**20**(12):855–9.
13. Mair EA, Parsons DS, Lally KP. Treatment of severe bronchomalacia with expanding endobronchial stents. *Arch Otolaryngol Head Neck Surg* 1990;**116**(9):1087–90.
14. Fabian KM. [The intra-prostatic "partial catheter" (urological spiral) (author's transl)]. *Urol Ausg A* 1980;**19**(4):236–8.
15. Oelke M, Bachmann A, Descazeaud A, et al. EAU guidelines on the treatment and follow-up of non-neurogenic male lower urinary tract symptoms including benign prostatic obstruction. *Eur Urol* 2013;**64**(1):118–40.
16. Peyton CC, Badlani GH. The management of prostatic obstruction with urethral stents. *Can J Urol* 2015;(22 Suppl. 1):75–81.
17. Vanderbrink BA, Rastinehad AR, Badlani GH. Prostatic stents for the treatment of benign prostatic hyperplasia. *Curr Opin Urol* 2007;**17**(1):1–6.
18. Oesterling JE, Kaplan SA, Epstein HB, Defalco AJ, Reddy PK, Chancellor MB. The North American experience with the UroLume endoprosthesis as a treatment for benign prostatic hyperplasia: long-term results. The North American UroLume Study Group. *Urology* 1994;**44**(3):353–62.
19. Armitage JN, Cathcart PJ, Rashidian A, De Nigris E, Emberton M, van der Meulen JH. Epithelializing stent for benign prostatic hyperplasia: a systematic review of the literature. *J Urol* 2007;**177**(5):1619–24.
20. Poulsen AL, Schou J, Ovesen H, Nordling J. Memokath: a second generation of intraprostatic spirals. *Br J Urol* 1993;**72**(3):331–4.
21. Lee G, Marathe S, Sabbagh S, Crisp J. Thermo-expandable intra-prostatic stent in the treatment of acute urinary retention in elderly patients with significant co-morbidities. *Int Urol Nephrol* 2005;**37**(3):501–4.
22. Armitage JN, Rashidian A, Cathcart PJ, Emberton M, van der Meulen JH. The thermo-expandable metallic stent for managing benign prostatic hyperplasia: a systematic review. *BJU Int* 2006;**98**(4):806–10.
23. Perry MJ, Roodhouse AJ, Gidlow AB, Spicer TG, Ellis BW. Thermo-expandable intraprostatic stents in bladder outlet obstruction: an 8-year study. *BJU Int* 2002;**90**(3):216–23.
24. Gottfried HW, Schlmers HP, Gschwend J, Brandle E, Hautmann RE. Thermosensitive stent (Memotherm) for the treatment of benign prostatic hyperplasia. *Arch Espanoles Urol* 1994;**47**(9):933–43. discussion 943–936.

25. Uchikoba T, Horiuchi K, Satoh M, et al. Urethral stent (Angiomed-Memotherm) implantation in high-risk patients with urinary retention. *Hinyokika Kiyo Acta Urolog Jpn* 2005;**51**(4):235–9.
26. Williams G, White R. Experience with the Memotherm permanently implanted prostatic stent. *Br J Urol* 1995;**76**(3):337–40.
27. Gesenberg A, Sintermann R. Management of benign prostatic hyperplasia in high risk patients: long-term experience with the Memotherm stent. *J Urol* 1998;**160**(1):72–6.
28. van Dijk MM, Mochtar CA, Wijkstra H, Laguna MP, de la Rosette JJ. Hourglass-shaped nitinol prostatic stent in treatment of patients with lower urinary tract symptoms due to bladder outlet obstruction. *Urology* 2005;**66**(4):845–9.
29. van Dijk MM, Mochtar CA, Wijkstra H, Laguna MP, de la Rosette JJ. The bell-shaped nitinol prostatic stent in the treatment of lower urinary tract symptoms: experience in 108 patients. *Eur Urol* 2006;**49**(2):353–9.
30. Shah DK, Kapoor R, Badlani GH. Experience with urethral stent explantation. *J Urol* 2003;**169**(4):1398–400.
31. Wilson TS, Lemack GE, Dmochowski RR. UroLume stents: lessons learned. *J Urol* 2002;**167**(6):2477–80.
32. Barber NJ, Roodhouse AJ, Rathenborg P, Nordling J, Ellis BW. Ease of removal of thermo-expandable prostate stents. *BJU Int* 2005;**96**(4):578–80.
33. Ala-Opas M, Talja M, Tiitinen J, Hellstrom P, Heikkinen A, Nurmi M. Prostakath in urinary outflow obstruction. *Ann Chir Gynaecol Suppl* 1993;**206**:14–8.
34. Thomas PJ, Britton JP, Harrison NW. The Prostakath stent: four years' experience. *Br J Urol* 1993;**71**(4):430–2.
35. Ozgur GK, Sivrikaya A, Bilen R, Biberoglu K, Gumele HR. The use of intraurethral prostatic spiral in high risk patients for surgery with benign prostatic hyperplasia. *Int Urol Nephrol* 1993;**25**(1):65–70.
36. Yachia D, Aridogan IA. Comparison between first-generation (fixed-caliber) and second-generation (self-expanding, large caliber) temporary prostatic stents. *Urol Int* 1996;**57**(3):165–9.
37. Yachia D, Aridogan IA. The use of a removable stent in patients with prostate cancer and obstruction. *J Urol* 1996;**155**(6):1956–8.
38. Corica AP, Larson BT, Sagaz A, Corica AG, Larson TR. A novel temporary prostatic stent for the relief of prostatic urethral obstruction. *BJU Int* 2004;**93**(3):346–8.
39. Henderson A, Laing RW, Langley SE. A Spanner in the works: the use of a new temporary urethral stent to relieve bladder outflow obstruction after prostate brachytherapy. *Brachytherapy* 2002;**1**(4):211–8.
40. Dineen MK, Shore ND, Lumerman JH, Saslawsky MJ, Corica AP. Use of a temporary prostatic stent after transurethral microwave thermotherapy reduced voiding symptoms and bother without exacerbating irritative symptoms. *Urology* 2008;**71**(5):873–7.
41. Yildiz G, Bahouth Z, Halachmi S, Meyer G, Nativ O, Moskovitz B. Allium TPS–a new prostatic stent for the treatment of patients with benign prostatic obstruction: the first report. *J Endourol Endourol Soc* 2016;**30**(3):319–22.
42. Kemppainen E, Talja M, Riihela M, Pohjonen T, Tormala P, Alfthan O. A bioresorbable urethral stent. An experimental study. *Urol Res* 1993;**21**(3):235–8.
43. Petas A, Talja M, Tammela TL, Taari K, Valimaa T, Tormala P. The biodegradable self-reinforced poly-DL-lactic acid spiral stent compared with a suprapubic catheter in the treatment of post-operative urinary retention after visual laser ablation of the prostate. *Br J Urol* 1997;**80**(3):439–43.

Prostatic Artery Embolization

20

Roger Valdivieso, Cristina Negrean, Pierre-Alain Hueber, Malek Meskawi, Khaled Ajib, Kevin C. Zorn
Centre Hospitalier de l'Université de Montréal (CHUM), Montréal, QC, Canada

Introduction

Benign prostatic hyperplasia (BPH) is the most common urologic pathology affecting aging men. An enlarged prostate can cause lower urinary tract symptoms (LUTS) by obstructing the bladder neck, thus increasing outflow resistance and increasing smooth muscle tone of the bladder. Moreover, epidemiologic and pathophysiologic associations have been established between LUTS/BPH and erectile dysfunction (ED).[1] 50% of men older than 60 years suffer from HBP.[2] If they experience LUTS, which affect their quality of life and daily activities, they are most often advised to initiate medical treatment. As stated by the American Urology Urological Association (AUA) and the European Association of Urology (EAU) guidelines, LUTS relief can be achieved medically by means of alpha-blockers or 5-alpha reductase inhibitors.[3,4] Combination of both medications has also demonstrated improved outcomes compared to either as monotherapy.[5] More recent studies demonstrate that a daily, low dose phosphodiesterase type 5 inhibitor such as a Tadalafil could be used as well to manage LUTS caused by BPH.[6]

Medical therapy is indicated for patients with moderate LUTS with no absolute surgical indications.[7] However, when medical control fails and/or complications occur, a surgical procedure can be attempted. Transurethral resection of the prostate (TURP), first described in 1932, has long been considered the unopposed gold standard for interventional BPH treatment of smaller prostates (≤60–80 cm^3). On the other hand, patients with larger prostates are usually offered open simple retropubic prostatectomy, as described by Millin et al. Because of high risk of complication and its invasive character, minimally invasive surgical techniques (MISTs) have emerged in recent years. MIST has proven to reduce costs and length of stay while equaling or even improving the functional outcomes of TURP.[8] The most common, approved minimally invasive techniques are intraprostatic stents, transurethral needle ablation, transurethral microwave therapy, and laser vaporization. So far, GreenLight XPS is the only technique that has shown superiority in treating LUTS compared to TURP from a coat versus benefit standpoint.[9] For these reasons and due to heterogenous worldwide access to such novel technologies, TURP remains the standard treatment option for most BPH cases.[10]

For over 30 years, the embolization of iliac arterial branches has been successfully used to manage severe prostatic hemorrhage secondary to prostate cancer or BPH.[11–14] The first published case in which prostatic arterial embolization (PAE) has been used as a therapeutic approach for BPH was in 2000 by DeMeritt et al.[13] However, it was only recently that PAE started to be offered as primary control treatment for LUTS caused by

A Comprehensive Guide to the Prostate. https://doi.org/10.1016/B978-0-12-811464-3.00020-9

BPH. First, its feasibility and safety was established from trials in dogs and pigs.[15–17] In humans, the first intentional treatment of BPH with PAE was done by Carnevale et al. in June 2008 and published in 2010.[18] The authors performed PAE for men with acute urinary retention (AUR) due to BPH who were waiting for surgery. They found that PAE reduced prostatic volume (PV) as well as postvoidal residual (PVR) urine volume. Moreover, better results have been achieved with bilateral PAE (39.1% reduction in PV with unilateral PAE vs. 47.8% with bilateral). Since then, multiple authors have published their small series of initial results, describing their efficacy, durability, and adverse event rates of PAE in symptomatic patients. Those are summarized in Table 20.4.

Prostatic Anatomy and Vascular Supply

Prostate Zonal Anatomy

The contemporary description of the prostate anatomy is based on the work of McNeal in 1981 that defined four basic anatomic regions on the analysis of 500 prostate specimens.[19] Each zone harbors specific embryologic and histologic features and is anatomically defined according to their spatial relationship to the prostatic urethra.

The prostatic urethra serves as a reference point and runs through the entire length of the prostate slightly closer to its anterior surface. A urethral crest projects inward from the posterior midline with prostatic sinuses on each side that drained the glandular periurethral ducts.[20]

At the level of its midportion, the urethra bends forward to form an anterior angle of approximately 30 degrees (varying from 0 to 90), which divides the prostatic urethra into a proximal (preprostatic) and a distal (prostatic) segment.[20]

In the proximal segment, circular smooth muscles are thickened to form the involuntary internal urethral (preprostatic) sphincter.[20] At the angle of the urethra the ducts of the transition zone (TZ) connect passing beneath the preprostatic sphincter to travel on its lateral and posterior side.[20] Duct development is limited in this region possibly because of their intimate relationship with the smooth muscle sphincter. Just distal to this angulation the urethral crest widens and protrudes from the posterior wall to form the verumontanum. At the apex of the verumontanum the utricle, a 6 mm müllerian vestigial remnant can be found flanked on each side of its orifice by the two small openings of the ejaculatory ducts. Around the openings of the ejaculatory ducts, ducts arise circumferentially the ducts of the central zone.[20]

Hence, the prostate has been divided into four zones according to the location of their ducts in the urethra, their histology, and their embryologic origin. This zonal anatomy is also consistent with the type of pathologic lesion that they develop.[18,19] These four zones include (1) the central zone (CZ) (2) the TZ (3) the peripheral zone (PZ), and (4) the anterior fibromuscular stroma (AFMS).

The Central Zone

The CZ represents 25% of the glandular prostate. Its ducts arise close to the ejaculatory duct orifices and follow these ducts proximally, branching laterally near the prostate base. Its lateral border fuses with the proximal PZ border, completing in continuity with the PZ, a full disc of secretory tissue oriented in a coronal plane. Marked histologic differences between central and PZs suggest important biologic differences.[19,21,22]

The Transition Zone

From endodermal origin (urogenital sinus) this zone surrounds the prostatic urethra proximal to the verumontanum. Normally, the TZ accounts for 5%–10% of the glandular tissue of the prostate and include two lateral lobes on either side.[19,21,22] A median lobe can also develop located between the two ejaculatory ducts and the urethra. As age increases, hyperplasia of the epithelial and stromal architecture can occur and give rise to BPH commonly encroaching on the urethral space and causing bladder outlet obstruction.

In addition, approximately 20% of adenocarcinomas of the prostate originate in this zone.[21] Cancers in the TZ show clinical features that are different from those shown by cancers in the PZ. Patients with TZ cancers have higher mean prostate-specific antigen (PSA) levels and higher tumor volume than patients with PZ cancers.[23,24] However, they tend to have lower Gleason scores and a more favorable prognosis.[24]

The Peripheral Zone

The PZ is of mesodermal (Wolffian duct) origin and constitutes the bulk (over 70%) of the prostate gland. It encloses and covers the posterior and lateral aspect of the gland. Posteriorly, the PZ lies against the rectum and is the region palpable by digital rectal examination (DRE). Its ducts radiate laterally from the urethra distal to the verumontanum. The majority (70%) of prostatic adenocarcinoma arises in this zone.[21] In addition this zone is the region most commonly affected by chronic prostatitis.[25]

The Anterior Fibromuscular Stroma

The AFMS forms a thick, nonglandular shield that extends from the bladder neck to the striated sphincter and covers the entire anterior surface of the prostate. This stroma is composed of smooth and striated muscles cells that secrete collagen and elastin fibers. The inner layer is fused with the underlying anterior portion of the gland and can be replaced by glandular tissue in adenomatous enlargement of the prostate. Its external surface is in continuity with the prostatic capsule, the anterior visceral fascia, and the anterior portion of the preprostatic sphincter.[19]

Vascular Supply of the Prostate

Precise knowledge of prostate vascular supply is *sine qua non* of selective PAE procedure. At the beginning of the century, the unequivocal existence of a specific prostate artery was subject of debate in the literature.[26] Since then, original works from cadaveric studies and in patients using refined angiography imaging have confirmed and well established the presence of the prostate arteries.[26,27] Nevertheless, to this day, lack of a strict nomenclature for prostate vascularization persists and may be explained both by its anatomical variations and the heterogeneous terminology used historically in the literature to describe prostatic vessels.[26,27]

According to cadaveric studies described by Clegg in 1955, the prostatic anatomy (PA) is a well-defined trunk with variable origins but a rather constant course.[26] Hence, the PA passes obliquely downward, forward, and medially along the anteroinferior surface of the bladder toward the prostate. At the junction of the bladder and the prostate PA typically divides into two equal size prostatic pedicles:

1. The prostate-vesical group or anterolateral pedicle that supplies the bladder neck and the periurethral portion of the gland;
2. The capsular group or posterolateral pedicle that supplies the caudal and the peripheric portion of the gland.

PAs are relatively small-sized diameters arteries varying between 1 and 2 mm that can have a common or separate origin. In addition, considerable tortuosity and anastomoses exist between the two groups.[27]

The Anterolateral Prostatic Anatomy Pedicle

These arteries branch off perpendicularly to the urethra at the prostatovesical junction and then travel in a caudal direction parallel to the urethra in the anterolateral quadrants typically at the 11 o'clock (on the right side) and the 1 o'clock (on the left side) positions, respectively.[26,27]

By supplying the TZ, including the periurethral portion this group of urethral arteries, represents the main blood supply of the adenomatous portion of the gland that is enlarged in BPH. Bouissou and Talazac who compared the cadaveric prostate vascularization in 100 pelvic halves comparing BPH and prostate cancer gland observed that this pedicle was frequently more developed in patients with BPH.[28] Hence, in patients with BPH treated with PAE, this anterolateral prostatic pedicle is generally the preferred artery to be embolized.[27] Not surprisingly, these vessels are also frequently encountered during TURP during which they potentially can be the source of significant bleeding (Fig. 20.1).[29]

The Posterolateral Prostatic Anatomy Pedicle

The second main branch of the PA gives rise to the capsular arteries. This group of arteries travels posterolaterally to the prostate usually at the 7 o'clock (on the right side) and the 5 o'clock (on the left side) positions and then penetrates the prostate perpendicularly supplying the peripheral glandular portion of the prostate. These capsular vessels, including arteries and veins, provide the scaffolding for the cavernous nerve forming together the neurovascular bundle.[26,27]

The posterolateral prostatic pedicle supplies most of the peripheral and caudal gland. In PAE for BPH treatment, this artery is only embolized when access into the anterolateral prostatic pedicle is unsuccessful. Sometimes the anterolateral prostatic pedicle can be a difficult artery to catheterize because of its tortuous trajectory and its tight angulations when it arises from the superior vesical artery. In addition, the presence of atherosclerotic changes close to the ostium of the superior vesical artery can

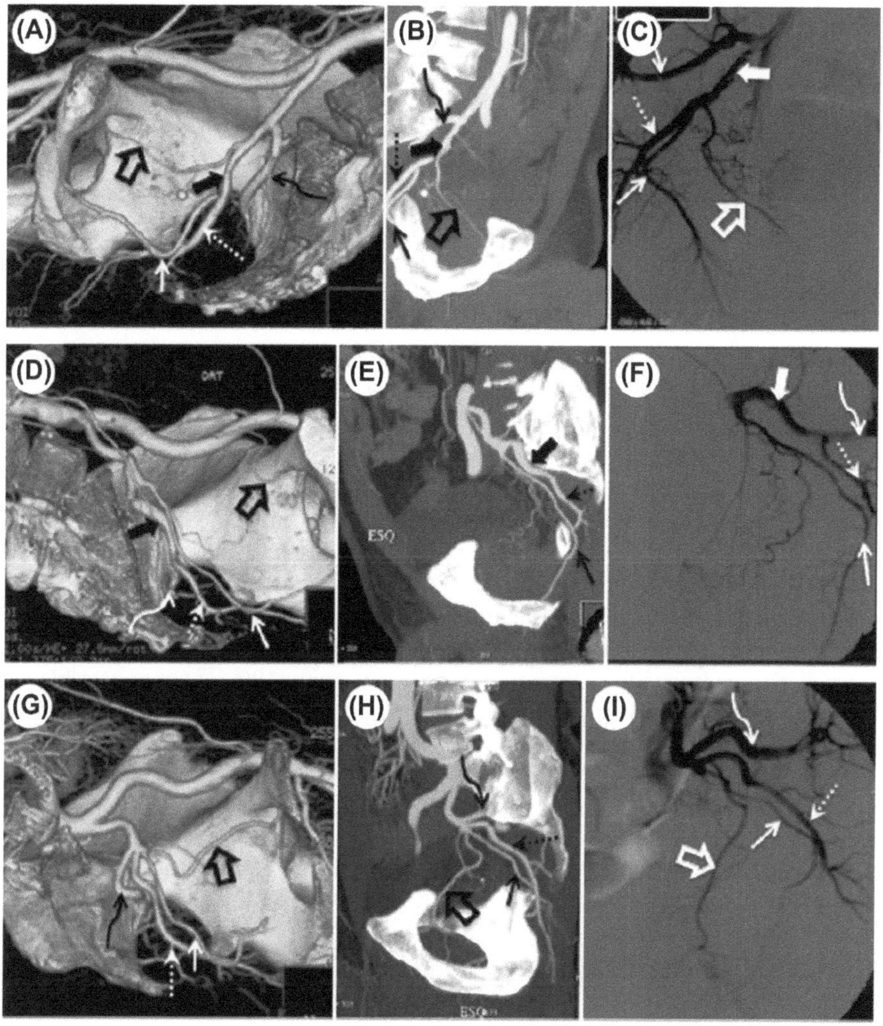

Figure 20.1 Computed tomography angiography (CTA) and digital subtraction arteriography (DSA) of the 3 main branching patterns of the internal iliac artery. (A) Type A bifurcation on the right CTA, 3D volume rendering, (B) type A bifurcation on the right CTA, sagittal MIP, and (C) type A bifurcation on the right, internal iliac artery DSA with right anterior oblique projection (351) and caudal–cranial angulation (101). In images (A–C) the posterior division gives the superior gluteal artery (*curved arrow*) and the anterior division is formed by the anterior common gluteal pudendal trunk (*solid arrow*) that bifurcates into internal pudendal artery (*straight arrow*) and inferior gluteal artery (*dotted arrow*). Obturator artery (*open arrow*) arising from the anterior common gluteal pudendal trunk. (D) Type B bifurcation on the left, CTA 3D volume rendering, (E) type B bifurcation on the left CTA, sagittal MIP, and (F) type B bifurcation on the left, internal iliac artery DSA with left anterior oblique projection (351) and caudal–cranial angulation (101). In images (D–F) the posterior division (*solid arrow*) gives the superior (*curved arrow*) and inferior (*dotted arrow*) gluteal arteries. The anterior division is formed by the internal pudendal artery (*straight arrow*). Obturator artery (*open arrow*) arising from the external iliac artery. (G) Type C bifurcation on the left, CTA 3D volume rendering, (H) type C bifurcation on the left, CTA sagittal MIP, and (I) type C bifurcation on the left, internal iliac artery DSA with left anterior oblique projection (351) and caudal–cranial angulation (101). All major vessels arise at the same time (superior [*curved arrow*] and inferior [*dotted arrow*] gluteal arteries, internal pudendal artery [*straight arrow*], and obturator artery [*open arrow*]). *MIP*, maximum intensity projection.

make selective catheterization even more difficult. In these cases, it can be sometimes possible to perform retrograde embolization of the other pedicle through intraprostatic interpedicular anastomoses.[27]

Anatomic Variations

As previously mentioned, variations in this region are not uncommon. First the site of origin of the PA is quite variable and has been described with different frequency of distribution. In a study by Ambrosio and colleagues on cadaveric specimens, the most common origin of the prostate artery was described from the inferior vesical artery (IVA) in 41.5%, internal pudendal 26.4%, and from the umbilical artery in 15.1% of the cases.[30] It can also come from obturator (5.7%), inferior gluteal (1.9%), and internal iliac (9.4%) arteries.[30] In contrast, in the most recent report by Bilhim et al. using computed tomography angiography (CTA), PAs most frequently arise from the internal pudendal artery (35%), from a common origin with the superior vesical artery (20%), from the common anterior gluteal–pudendal trunk (15%), from the obturator artery (10%), or from a the common prostatorectal trunk (10%).[27] Other possible but rare origins included from the inferior gluteal artery, superior gluteal artery, or from an accessory pudendal artery (10%).[27] In the same study, they have confirmed that two prostatic pedicles may arise from the same artery in patients with only 1 PA (found in 60% of pelvic sides) or may arise independently in patients with two independent PAs (found in 40% of pelvic sides).[27]

In the presence of two separate prostatic vascular pedicles, there is one anterolateral prostatic pedicle with a superior or proximal origin usually from the common anterior gluteal–pudendal trunk close to or with a common origin with the superior vesical artery. When there is a common arterial trunk with the vesical and the anterolateral PAs, it is advisable to advance the microcatheter tip distally to the vesical arteries.[27]

The posterolateral prostatic pedicle has an inferior or distal origin generally from the internal pudendal or obturator artery above the sciatic notch. At this location, it can have a close relationship with the rectal or anal branches. It is not uncommon to find a common trunk between the posterolateral vessels and the middle rectal artery that has to be avoided during catheterization.[27]

When only one PA is present, it usually arises independently from the vesical arteries (from the internal pudendal, obturator, or common anterior gluteal–pudendal trunk) and after a variable length it bifurcates into the anterolateral and posterolateral prostatic pedicles. In these cases, embolization of both prostatic pedicles can be performed with the tip of the catheter before the PA bifurcation (Fig. 20.2).[27]

Prostatic Anatomy Imaging

As mentioned above, PAs are relatively small vessels (1–2 mm diameter) with frequent variations and thus can be sometimes difficult to identify. Hence visualization of prostatic arteries using preprocedural imaging is advised to plan a successful intervention

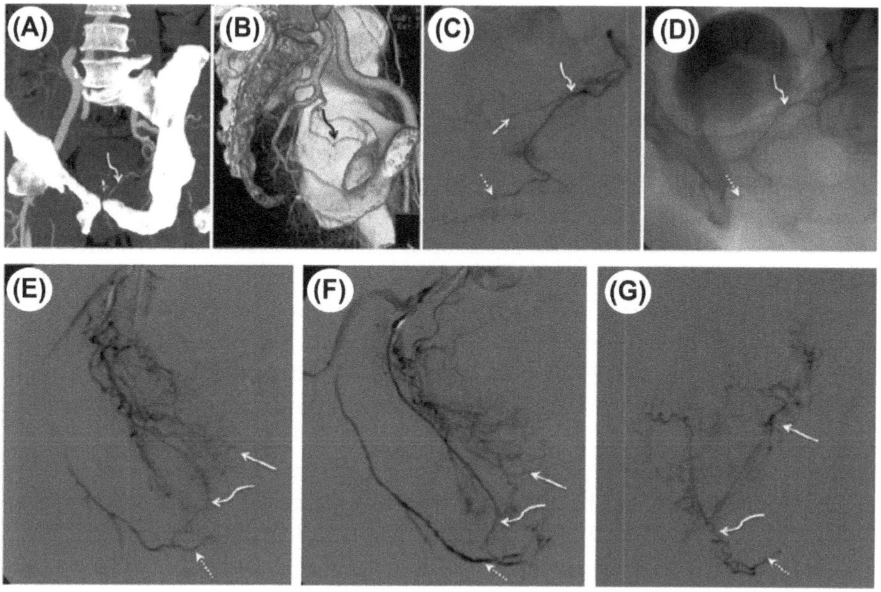

Figure 20.2 Accessory or aberrant pudendal arteries (APAs)—curved arrows. Lateral APA on the left pelvic side: (A) Computed tomography angiography (CTA) coronal MIP, (B) CTA 3D volume rendering, (C) selective digital subtraction arteriography (DSA) with left anterior oblique projection (351) and caudal–cranial angulation (101), and without subtraction (D). Note the trajectory of the APA behind the pubic bone, near the prostate surface and in close relationship with the prostatic artery (*straight arrow*), terminating as the main arterial supply to the penis (*dotted arrow*). APAs (*curved arrows*) representing large sized anastomoses between the prostatic arteries (*straight arrows*) and the termination of the internal pudendal artery (*dotted arrows*), representing collateral blood supply to the penis. (E) and (F) DSA of the right PA with right anterior oblique projection (351) and caudal–cranial angulation (101) in 2 patients, (G) DSA of the left PA with left anterior oblique projection (351) and caudal–cranial angulation (101) in another patient (*Curved arrow* ¼ _APA, *straight arrow* ¼ _prostatic artery, *dotted arrow* ¼ _penile artery). *MIP*, maximum intensity projection.

and to prevent complications resulting from untargeted embolization of the adjacent organs such as the bladder, rectum, and penis. According to Bilhim T et al., preoperative CTA can depict the pelvic arterial anatomy with excellent correlation with digital subtraction arteriography (DSA) findings per procedure and therefore be used to guide PAE.[27,31]

Patient Selection

Preprocedural clinical evaluation and patient selection for PAE are paramount to improve technical and clinical results. For this reason rigorous inclusion and exclusion criteria are applied. These criteria may vary from one group performing the procedure to the other. Here we describe the main criteria used by the leading groups performing PAE.[20,32–34]

Patient Evaluation

Initially, the urologist performs a clinical evaluation to determine the nature and severity of the disease, excluding other sources of BOO. The evaluation should include the assessment of symptoms using validated questionnaires such as the International Prostate Symptoms Symptom Score (IPSS) and quality of life (QoL) to grade severity. During physical exam, a DRE is performed to estimate prostate volume and detect concomitant cancer. Unfortunately, the accuracy of actual prostate volume and DRE interpretation becomes highly unreliable for prostates >60cc. For men with suspected larger prostates, a transrectal ultrasound (TRUS) can be performed to evaluate prostate volume more precisely. Bienz et al. have previously reported a correlation between accuracy of TRUS and prostate volume.[35] They have found a mean absolute percentage of error of 15% for prostates bigger than 80g—which yields a far more reliable result than DRE. PSA should also be reviewed to detect the probability of disease progression (high-risk patients) and of prostatic cancer, guiding the eventual need for prostatic biopsy.

Uroflow studies can also be performed before and after PAE to examine maximum flow rate (Q_{max}) and PVR. If the Q_{max} is very low (<6–8mL/s), a urethral stricture is to be excluded by performing an urethrocystoscopy. Carnevale et al.[33] include a transrectal or suprapubic ultrasound and a pelvic MRI to establish a documented baseline for each patient. They also, also include a urodynamic evaluation despite its invasive nature to exclude patients that experience infravesical obstruction associated with other bladder disorders and detrusor hypocontractility.

Finally, if the patient is considered for PAE, a computed tomographic angiography (CTA) is performed to study the patient's pelvic arterial anatomy. In addition, an MRI and/or a transrectal/suprapubic ultrasound may also be performed to establish a baseline PV.

Inclusion Criteria

According to most recent data, authors suggest including male patients aged >40 years with symptomatic BPH refractory to medical treatment for at least 6 months.[32] Patients should have moderate to severe LUTS, described as IPSS > 12, Qmax < 15 mL/s, or QoL > 3.[36–38] Gao et al.[39] have included patients with a prostate volume starting from 20 mL. Also, they have used a different threshold for LUTS severity (IPSS ≥ 7) to consider PAE. Furthermore, other authors have included men who are intolerant to medical modalities of treatment.[33] There is a controversy regarding patients with AUR. Lately, it has been considered to be more of an exclusion criterion.[40] Conditions that are refractory to medical therapy are also considered.

Exclusion Criteria

The exclusion criteria include malignancy (based on dubious TRUS or DRE with positive biopsy, or PSA > 4 ng/mL with positive biopsy), bladder abnormalities such as large diverticula or large stones, bladder atonia, and neurogenic bladder.[39] Unregulated

Table 20.1 Summary of selection criteria for prostatic arterial embolization published so far

Inclusion criteria
1. Men >40 years with BPH causing LUTS refractory to medical Tx >6 mo
2. Moderated to severe LUTS:
IPSS ≥12
Peak urinary flow <15 mL/s
QoL ≥3
3. PSA <4 ng/mL
4. PSA >4 ng/mL with negative biopsy
5. Prostate volume ≥20 mL
Exclusion criteria
1. Chronic renal failure
2. Diabetes mellitus
3. Severe atherosclerosis
4. Severe tortuosity in internal iliac arteries
5. Unregulated coagulation parameters
6. Previous prostate surgery
7. Prostatic cancer
8. Bladder abnormalities:
Large bladder diverticula (>5 cm)
Large bladder stones (>2 cm)
Active UTI
Neurogenic bladder
Bladder atonia
Bladder neck contracture
Sphincter decompensation
Episode of AUR in last 4 weeks
9. Urethral stricture
10. Patients with indwelling urinary catheter
11. Life expectancy <2 years

AUR, acute urinary retention; *IPSS*, international prostate symptom score; *LUTS*, lower urinary tract symptoms; *PSA*, prostatic-specific antigen; *QoL*, quality of life; *UTI*, urinary tract infection.

coagulation parameters, chronic renal failure, tortuosity, and advanced atherosclerosis of iliac or prostatic arteries on preprocedural CTA are also important exclusion factors. Li et al.[40] as well as some other studies excluded patients who had an episode of AUR in last 4 weeks or active urinary tract infection. Respecting these criteria, Perreira et al. have noted that only a third of patients initially seen on consultation for refractory LUTS, actually qualifies for PAE procedure.[32]

See Table 20.1 for a complete list of inclusion and exclusion criteria.

Embolization Technique

The technique requires a well-trained interventional radiologist beyond the learning curve because of the complexity of the vascular anatomy of the prostate as well as the possible complications that could arise in more difficult cases. It is very important to properly identify the vasculature of the prostate and the variants that may occur among patients. The technique may vary from one group to the other. Here we present the so far published PAE techniques.[20,30,31,38,39] Informed consent should be obtained after discussing all potential complications. These will be detailed further below.

Preparation

The intervention is done under local anesthesia and, most often, on an outpatient basis in the interventional radiology suite. An oral fluoroquinolone is usually administered prior the procedure as antibioprophylaxis. The antibiotic may be continued 7–10 days postprocedure depending on the authors. Nonopioid analgesia is provided as well as nonsteroidal antiinflammatory medications before and after PAE.

A Foley catheter is inserted prior the procedure and the balloon is inflated with a mixture solution of iodine contrast and normal saline. This helps provide good orientation to the prostate site and related pelvic structure during the procedure.[33]

Identification and Catheterization of Prostatic Arteries

Embolization is performed via either unilateral femoral approach, most commonly, or sometimes via bilateral approach. Thus, femoral pulses should be examined before to choose the best puncture site. The angiography is started by introducing usually a 4-Fr catheter via the right common femoral artery to catheterize the left (or right) internal iliac artery. Then using a pump injection (20 mL; 10 mL/s) the iliac vessels and the prostatic arteries are identified and evaluated. To better assess the blood supply to the prostate, a 5-French vertebral catheter (12 mL; 4 mL/s) is advanced at the common internal iliac trunk and a selective DSA of the internal iliac artery is performed using the ipsilateral oblique projection (30–35 degrees) and caudal–cranial angulation (10 degrees) for better visualization. This way omission of any artery arising from the anterior and posterior division of the iliac is avoided. At this point, cone-beam computer tomography with 4–6 s delay can be performed to evaluate for possible nontarget embolization. If such sites are identified, the catheter is advanced a little further to ensure selective embolization of the PAs.[40]

Carnevale et al. prefer to use the 25–55 degrees ipsilateral oblique view as well as a caudal view (10–20 degrees) to better identify the IVA and all possible accessory branches to the prostate. These views combined with the Foley's balloon filled with contrast help better understand the anatomy and all the branches that irrigate the prostatic gland. Under this ipsilateral oblique perspective, catheterization of the IVA is performed with a small (<2.3 Fr) hydrophilic microcatheter and 2–3 mL of contrast are injected manually. At this point the posterior–anterior view is used as it helps to identify some contralateral prostate lobe branches, and the parenchymal phase. The microcatheter is then introduced deeper into the IVA at the ostium of the prostatic arteries and embolization is performed under direct visualization. If vessel spasm occurred, nitroglycerin (200–300 μg) was injected intraarterially.

Embolization

Numerous embolic agents can be used for embolization of prostatic arteries. Carnevale et al. prefer to use 100–300 μm and 300–500 μm microspheres (Embosphere). A recent prospective study by Wang et al.[36] showed that using 50 μm microspheres for distal intraprostatic embolization and 100 μm for proximal embolization results in better outcomes since those small particles can penetrate in the vessels irrigating the periurethral region of the prostate. A study by Li et al. shows that this technique yields greater ischemia and infarction of the prostate and thus lower risk of failure and better symptom control.[40] The embolization material (2 mL) is mixed in a 20 mL 50–50 saline/contrast solution adding up to 22 mL. Pisco et al.[41] use an 80 mL solution of 1 mL polyvinyl alcohol (PVA) particles and 50–50 contrast saline solution. Whereas Li et al.[40] used 1 mL of PVA in 40 mL solution of nonionic contrast medium. This solution is injected slowly under direct fluoroscopy. It is estimated that embolization requires 10–15 min to reach endpoint. Wang et al.[36] injected nitroglycerin (200–300 μg) intraarterially per procedure to increase vessel's size and thus achieve a more effective embolization. Some authors would change the particle size per procedure depending on the level of pain experienced by the patient to avoid untargeted embolization.[41] After 3–5 min of injecting the embolization product, a flush of normal saline should be injected to pack the microspheres inside the prostate. If continued forward flow is observed, more embolic agent may be injected. The endpoint is considered to be total stasis in the prostatic vessels with interruption of blood flow and prostatic gland opacification. In general, a single syringe of 2 mL of microspheres is sufficient for bilateral PAE. After the endpoint is reached, the microcatheter is retired to the origin of the IVA and a manual injection of contrast is performed for final control and to look for additional prostatic branches. Embolization is later performed on the contralateral side using the same technique. In some instances, microcoil deployment can be used as an alternative mechanism for embolization.[42]

Postprocedure Management

Postoperative patients are to remain 4–6 h without moving the punctured leg to avoid bleeding complications. If a vascular closure device is used, resting time can be reduced. Appropriate hydration is administered as well as antibioprophylaxie to prevent infection.[36,39] The Foley catheter is left in place to help voiding without straining, which helps to reduce the risks of puncture site complications. The catheter is removed 2–4 h postprocedure. If the patient is not in AUR, then the patient is discharged. Some authors describe a longer stay, up to 5–7 days, if patients have heavy comorbidities.

Patients with AUR and chronic use of indwelling catheter are asked to come back 1 week later and trials are attempted in 1-week intervals. A clinical failure is determined if the patient cannot spontaneously void in 1 month.

The most common symptom after the procedure is dysuria and frequent urination and could last for 3–5 days. Patients are asked to stop prostatic medications just after PAE. All patients are seen in outpatient consult 1 month after the procedure.

Technical and Clinical Success

Technical success is achieved when main bilateral prostatic arteries are embolized and much prostatic ischemia is achieved. The main purpose of PAE is to achieve prostatic ischemia as it has been established that better long-term clinical and urodynamic results are correlated with prostate ischemia.[33] Clinical success is defined as either the removal of the Foley catheter in patients with AUR, improvement of symptoms as seen in an improvement of IPSS and QoL questionnaires and no sexual disorders or major adverse events after the procedure.

Published Outcomes

Even though more than 250 studies have been published so far, very few of them show high level of evidence regarding PAE technique. A fairly large amount of studies do not include all follow-up parameters required to make scientific decisions. Moreover, the longest follow-up time is only 2 years. Therefore, making it even harder to draw consistent conclusions on long-term outcomes (Table 20.2).

Since the initial reports of PAE reporting, more meaningful data have appeared. A recent metaanalysis published in November 2016 by Feng et al. reviewed 20 articles on PAE technique as treatment option for LUTS due to BPH. Their analysis showed some modest but significant clinical improvement in patients who underwent PAE.[43] Indeed, on average, males see 13.21 points amelioration on their IPSS scale after PAE. PV decreases from 20.07 up to 35.94 mL with an average of 28 mL. Consequently, there has been observed a statistically significant decrease in postvoid residual volume (mean of −69.85 mL), in PSA levels (mean - 1.35 ng/mL), and a statistically significant increase in Qmax (mean + 5.28 mL/s). The erectile function score had no significant decrease at follow-up. However, patients' quality of life seems to show only limited improvement, by 2.34 points on average.[43] For detailed results on IPSS, QoL, PV, and Qmax overtime please refer to Table 20.3. It includes data from seven big articles on PAE.

Complications of PAE seem to be relatively few. Among those patients who experience some inconfort after the procedure, authors have found a collection of symptoms known as "post-PAE syndrome." It has been often described[42,43,45] but the authors do not agree on its exact definition. Indeed, post-PAE syndrome might include retropubic or perineal pain, nausea, vomiting, urethral burning, fever, and/or small amount of hematuria or rectorragia. Some more rare but potentially serious complications are urinary tract infection (UTI) (3.10%), balanitis/balanoprostatitis (0.30%), bladder ischemia (0.20%), retrograde ejaculation (2.30%), and inguinal hematoma (1.60%).[43] For a complete list of complications, please see Table 20.4.

Recent published data show clinical failure rate of PAE as being somewhere below 10%. Indeed, Wang et al.[36] reported a failure rate of 6.7%. Whereas Li et al.[32] had a 9% failure rate. Same results (9.4%) were found by Gao et al.[39] In all three studies, the majority of patients who experienced relapse in LUTS symptoms were patients who had only unilateral embolization. MRI of those patients actually shows regrowth of the prostate[34]—which is probably the cause of failure. Note that, long-term failure rate still remains unknown as the maximum follow-up is only 2 years. Abt et al. are now

Table 20.2 Published prostatic arterial embolization studies

Author	Year	Type	Material (μm)	Number of subjects	AUR (%)	Age (mean)	Mean follow-up (months)	Time of embolization mean (min)	Technical success (%)	Clinical success (%)
Pisco et al.[23]	2013	Prospective	200, PVA nonspherical	255	13%	65.5	10	73	98%	72%
Antunes et al.[34]	2013	Prospective	300–500, microspheres	11	100%	68.5	N/A	197	100%	91%
Wang et al.[36]	2015	Prospective	50, 100 PVA	64	31.7%	74.5	18	115	93.8%	95%
Kurbatov et al.[38]	2014	Prospective	300–500 trisacryl microspheres	88	N/A	66.4	N/A	84	N/A	N/A
Gao et al.[39]	2014	Prospective	355–500 PVA microspheres	114	25.9%	N/A	22.5	90	94.7%	90.6%
Li et al.[40]	2015	Prospective	50, 100 PVA	24	32%	74.5	14	115	92%	91%
Rio Tinto et al.[44]	2012	Retrospective	100–200 PVA, nonspherical	103	15%	66.8	N/A	83	97%	89%

AUR, acute urinary retention; *PVA*, polyvinyl alcohol.

Table 20.3 Published follow-up outcomes

Author	Mean IPSS						Mean QOL					
	0	1	3	6	12	24	0	1	3	6	12	24
Pisco et al.[26]	24	12.2	11	11.5	10.4	9	4.4	2.5	2.2	2.3	2.0	1.8
Antunes et al.[34]	N/A	7.1	N/A	N/A	2.8	N/A	N/A	N/A	N/A	N/A	N/A	N/A
Wang et al.[36]	27.0	10.0	7.5	7.0	7.0	8.0	5.5.	2.0	2.5	2.5	2.0	2.0
Kurbatov et al.[38]	23.98	N/A	12.2	11.35	10.4	N/A	5.1	N/A	2.8	2.78	2.2	N/A
Gao et al.[39]	22.8	19.2	15.6	12.8	10.9	8.7	4.8	3.7	2.9	2.2	1.9	1.6
Li et al.[40]	27.0	12.0	7.0	8.0	7.5	N/A	4.50	2.5	2.0	2.0	2.0	N/A
Rio Tinto et al.[44]	22.8	11.7	10.3	10.8	11.2	9.3	4.1	2.3	2.0	2.0	2.0	2.0

IPSS, international prostatic symptoms score; Qmax, maximal urinary flow rate; *QOL*, quality of life.

Table 20.4 Published outcomes of prostatic arterial embolization: efficacy and complication rate

Author	Efficacy (%)		Failure rate (%)			
	Technical[a] success	Clinical[b] success	Clinical[c]	Urethral burning	Hematuria	Hematospermia
Pisco et al.[23]	98%	72%	28%	9.2%	5.6%	4.0%
Wang et al.[36]	93.8%	95%	6.7%	18.3%	10.0%	10.0%
Russo et al.[37]	N/A	N/A	N/A	6.25%	0%	1.25%
Gao et al.[39]	94.7%	90.6%	9.4%	N/A	0%	N/A
Li et al.[40]	92%	91%	9.0%	36%	14%	9%
Feng et al.[43]	N/A	N/A	N/A	9.40%	9.00%	5.20%

AUR, acute urinary retention; *UTI*, urinary tract infection.

[a]Technical success is defined in most studies as unilateral or bilateral PAE as shown by selective angiography carried out at completion of the procedure.

[b]Clinical success is defined in most studies as improvement in symptoms (IPSS reduction ≥25% and total IPSS score <18 points; QoL reduction >1 point or total score ≤4; Qmax improvement by ≥2.5 mL/s, and a total score ≥7 mL/s).

[c]Clinical failure is defined in most studies as persisting severe symptoms (IPSS reduction <25% and/or IPSS ≥18; QoL reduction ≤1 point and/or total QoL >4; Qmax improvement by <2.5 mL/s or total Qmax <7 mL/s) or additional treatments needed. However, the authors did not report a precise reintervention rate. **Please note that while those are common definitions among the studies, they lack standardized BPH data to confirm success rates**.

Mean prostate volume (mL)						Mean Qmax (mL/s)					
0	1	3	6	12	24	0	1	3	6	12	24
83.5	66.8	68.3	66.6	69.9	72	9.2	11.9	12.4	12	12.8	13.9
69.7	N/A	N/A	N/A	N/A	N/A	0	11.9	N/A	N/A	N/A	N/A
121.0	105.5	70.5	71.0	69.5	71.5	7.0	12.0	13.5	12.0	14.0	13.0
129.3	N/A	87.3	69.5	71.2	N/A	7.28	N/A	14.95	16.2	16.9	N/A
64.7	50.1	43.4	36.3	35.6	N/A	7.8	13.1	17.3	21.5	22.1	21.5
110	100	68.0	67.0	69.0	N/A	6.00	12.0	13.0	13.0	12.0	N/A
88	68.7	67	67.8	68	76	8.7	11.6	12.7	12.6	12.9	14.4

Complication rate (%)						
Retrograde ejaculation	Rectal bleeding	AUR	UTI	Balanitis	Bladder ischemia	Inguinal hematoma
N/A	2.4%	2.4%	1.2%	1.6%	0.4%	N/A
N/A	8.3%	31.7%	N/A	N/A	N/A	N/A
N/A	N/A	N/A	1.25%	N/A	N/A	N/A
N/A	N/A	25.9%	1.9%	N/A	N/A	N/A
N/A 2.30%	14% 3.90%	32% 7.00%	N/A 3.10%	N/A 0.30%	N/A 0.20%	N/A 1.60%

conducting a randomized noninferiority trial comparing PAE versus TURP results over a period of 5 years.[2] Hopefully, this study will bring more clarity in terms of efficacy, failure rate, complications, safety and indications of PAE.

Complications

Similar to other visceral embolization procedures PAE is associated with common symptoms such as nausea, vomiting, and fever in the absence of infection. Some more specific symptoms include urethral burning, periprostatic or pelvic pain, and non-significant rectorragie and/or hematuria. This constellation of symptoms is referred to as the "post-PAE syndrome" by Carnevale et al.[33] This syndrome is believed to be the result of the migration of PVA particles through the anastomosis between the prostatic and the neighboring arteries.[46] It is usually self-limiting and requires no intervention. Antiinflammatory drugs are thought to be sufficient to prevent those symptoms.[42] Pisco et al.[23] reported in their latest prospective study of 250 patients the following adverse events: urinary tract infections requiring antibiotics (7.6%), transient hematuria (5.6%), transient hematospermia (0.4%), prostatitis (1.6%), and inguinal hematoma (7%). Six patients (0.02%) that underwent PAE had AUR and a temporary bladder catheter had to be placed for a couple of hours. One major complication was reported. One patient developed bladder wall ischemia resulting surgical management to remove the necrotic tissue. This was secondary to untargeted embolization of the bladder. They believe that this complication was due to a proximal and aggressive embolization of both prostatic and vesical arteries. In a prospective study of 11 patients with urinary retention due to BPH, Antunes et al. reported no major complications after PAE.[34] However, they observed that nine patients reported mild and transitory pain, three patients presented minimal rectal bleeding, and two had diarrhea lasting 24 h. One patient had a single episode of hematuria. No groups have reported sexual dysfunction as a complication post-PAE.

Another possible complication with PAE is radiation. Indeed, this technique can require a potentially significant fluoroscopy time lasting up to 30 min. Lately, newer technology and low-dose protocols seem to reduce the radiation burden. Gao et al. reported a mean radiation dose of 11305.1 cGy/cm^2—which is slightly higher than a gastrointestinal barium meal.[39] Despite that, further studies regarding the long-term outcomes of fluoroscopy are needed to assess safety of PAE technique.

Note that contrast nephropathy is not listed as a complication from the available studies, as renal failure is an exclusion criteria for patient selection and an intensive hydration is part of the procedure protocol. However, it is not to be overseen as non-negligible amounts of contrast material is used during the procedure.

Discussion

PAE technique has emerged as an interesting approach in quest for a nonsurgical alternative to standard BPH treatment. It is especially compelling for patients who do not respond to medical treatment or who are not well suited for surgery and minimally invasive interventions.

Lately, the research in this field has expended. Nevertheless, the data seem to remain preliminary. Most of the clinical data on the procedure still come from three cohorts (the largest including 255 patients) treated by Pisco JM and Bilhim T et al. in Lisbon (Portugal).[23] Other smaller studies have been published since, including a 114 patients cohort by Gao et al. in Chongqing (China)[39] and another cohort of 88 patients by Kurbatov et al. in Moscow, Russia.[38]

This chapter reviewed the major paperworks on PAE and presents a concise résumé of the data. We found that PAE is relatively safe intervention if performed by experienced radiologists. Indeed, prostatic vascularization is highly variable demanding precise knowledge of prostate vascular supply, thus making it a relatively complex procedure. Nevertheless, PAE is associated with low morbidity and no irritative urinary symptoms or sexual dysfunction.[45] After the procedure some of the patients experienced a constellation of symptoms described as "post-PAE syndrome": nausea, vomiting, fever in the absence of infection, and pelvic pain. This syndrome can last up to 2–3 days after PAE. Also, a transitory increase in PSA was noted in the first 24 h after PAE, which resolves completely by the time of first follow-up visit at 1 month. It has been reported to even decrease by up to 27% after 12 months.[45]

In most of cases, only minor complications such as UTI, transient hematuria, hematospermia, and inguinal hematomas have been reported.[41] However, the risk of complications secondary to off-target embolization to adjacent organs is present and can potentially be serious. For example, one patient from Pisco group had bladder wall ischemia that required surgical management.[41] Similarly we found one case report describing ischemic rectitis as a result of rectal nontarget embolization.[47] Understandably, the use of preprocedural CTA is key for both planning a successful intervention and limiting potentially serious complications.

Feng et al.[43] suggest in their 20 article metaanalysis that PAE can offer a statistically significant decrease in the PV and consequently, in postvoid residual volume, in PSA level and in symptoms. Moreover, PAE has no significant impact on erectile function score. Therefore, patients' quality of life as well as their IPSS score can be notably improved.

Lately, trials comparing PAE with TURP and with open prostatectomy (OP) have been published. Russo et al. showed that OP has a slightly higher intra- and postoperative complication rate, but this difference was not statistically significant. However, they noticed a 2.67-fold increase in risk of 1 year persistent symptoms after PAE, which might need reintervention.[37] Similar findings were described when comparing PAE with TURP. It seems that main advantage of PAE results from lower intra- and postoperative complication rate as well as from possibility to relieve inoperable patients from LUTS.[39]

These results suggest that PAE is at least effective and maybe more effective than medical therapy but most likely inferior to most minimally invasive procedures, including TUMT or TUNA. Older patients can definitely benefit from this minimally invasive technique with low risk of complication. Despite those advantages, some limitations have to be considered. Several papers raise an important concern on the long-term efficacy of PAE. As mentioned above, a small proportion of patients experience relapse in LUTS symptoms. This is probably due to de novo prostate growth as suggests MRI.[34] Prostate regrowth can be explained by its revascularization from accessory arteries or from the contralateral prostatic artery if unilateral embolization was performed.[37] In this case a reintervention might be needed.

It is worth mentioning that in general about 10% of patients eligible for PAE are finally excluded because of advanced atherosclerotic disease or tortuosity of the iliac and prostatic arteries. In addition, the procedure cannot be completed in another 3%–5% of patients for the same reason.[41] This represents a total of 10%–15% of patients initially eligible for all procedures but in which finally PAE is not technically possible. Likely, this proportion is significantly higher in patients who are not suited for surgery. This population is typically older and has a higher prevalence of significant atherosclerotic disease. Therefore, atherosclerosis represents another potential limitation for candidates who prefer PAE.

In conclusion, PAE is an emerging technique gaining in popularity among research physicians. With the current body of literature, PAE could be recommended for BPH management in patients with multiple comorbidities, refractory to medical treatment and who have a have a high operative risk. Indeed, PAE can be performed as an outpatient procedure. It has low complication rate and no upper limit for prostatic size. Moreover, patients with prostate volume >80 mL are especially compelling candidates as PAE is believed to work based on prostate volume reduction, which will be more significant in patients with big prostates. However, the risk of complication and failure of PAE should be weighed against its benefits. So far, no significant difference was noted in short-term outcomes of PAE versus TURP. Long-term outcomes of PAE remain unknown still. Finally, despite interesting preliminary results, if PAE wants to compete with the contemporary minimally invasive procedures currently available, it still has to prove equivalency or less adverse effects in head-to-head comparison trials. Hopefully, the upcoming results of a 5-year noninferiority trial comparing PAE with TURP by Abt et al.[2] will bring more light on the viability of the technique as well as on the long-term outcomes of PAE and thus improving its candidacy for FDA approval or for mention in any current professional practice guidelines such as AUA or EAU guidelines. For now, PAE is not a standard of care in BPH management. Moreover, most of the published data lack long-term durability, objective standardized questionnaires as well as urodynamics studies.

References

1. Andersson KE, et al. Tadalafil for the treatment of lower urinary tract symptoms secondary to benign prostatic hyperplasia: pathophysiology and mechanism(s) of action. *Neurourol Urodyn* 2011;**30**(3):292–301.
2. Abt D, Mordasini L, Hechelhammer L, Kessler TM, Schmid HP, Engeler DS. Prostatic artery embolization versus conventional TUR-P in the treatment of benign prostatic hyperplasia: protocol for a prospective randomized non-inferiority trial. *BMC Urol* 2014;**14**:94.
3. Madersbacher S, et al. EAU 2004 guidelines on assessment, therapy and follow-up of men with lower urinary tract symptoms suggestive of benign prostatic obstruction (BPH guidelines). *Eur Urol* 2004;**46**(5):547–54.
4. McVary KT, et al. Update on AUA guideline on the management of benign prostatic hyperplasia. *J Urol* 2011;**185**(5):1793–803.

5. Montorsi F, Roehrborn C, Garcia-Penit J, Borre M, Roeleveld TA, Alimi JC, et al. The effects of dutasteride or tamsulosin alone and in combination on storage and voiding symptoms in men with lower urinary tract symptoms (LUTS) and benign prostatic hyperplasia (BPH): 4-year data from the combination of avodart and tamsulosin (CombAT) study. *BJU Int* 2011;**107**(9):1426–31.
6. Oelke M, et al. Monotherapy with tadalafil or tamsulosin similarly improved lower urinary tract symptoms suggestive of benign prostatic hyperplasia in an international, randomised, parallel, placebo-controlled clinical trial. *Eur Urol* 2012;**61**(5):917–25.
7. Burnett AL, Wein AJ. Benign prostatic hyperplasia in primary care: what you need to know. *J Urol* 2006;**175**(3 Pt 2):S19–24.
8. Baazeem A, Elhilali MM. Surgical management of benign prostatic hyperplasia: current evidence. *Nat Clin Pract Urol* 2008;**5**(10):540–9.
9. Thomas JA, Tubaro A, Barber N, d'Ancona F, Muir G, Witzsch U, et al. A multicenter randomized noninferiority trial comparing GreenLight-XPS laser vaporization of the prostate and transurethral resection of the prostate for the treatment of benign prostatic obstruction: two-yr outcomes of the GOLIATH study. *Eur Urol* 2016;**69**(1):94–102.
10. Varkarakis J, Bartsch G, Horninger W. Long-term morbidity and mortality of transurethral prostatectomy: a 10-year follow-up. *Prostate* 2004;**58**(3):248–51.
11. Mitchell ME, et al. Control of massive prostatic bleeding with angiographic techniques. *J Urol* 1976;**115**(6):692–5.
12. Bischoff W, Goerttler U. Successful intra-arterial embolization of bleeding carcinoma of the prostate (author's transl). *Urol A* 1977;**16**(2):99–102.
13. DeMeritt JS, et al. Relief of benign prostatic hyperplasia-related bladder outlet obstruction after transarterial polyvinyl alcohol prostate embolization. *J Vasc Interv Radiol* 2000;**11**(6):767–70.
14. Barbieri A, et al. Massive hematuria after transurethral resection of the prostate: management by intra-arterial embolization. *Urol Int* 2002;**69**(4):318–20.
15. Jeon GS, et al. The effect of transarterial prostate embolization in hormone-induced benign prostatic hyperplasia in dogs: a pilot study. *J Vasc Interv Radiol* 2009;**20**(3):384–90.
16. Sun F, et al. Transarterial prostatic embolization: initial experience in a canine model. *AJR Am J Roentgenol* 2011;**197**(2):495–501.
17. Sun F, et al. Benign prostatic hyperplasia: transcatheter arterial embolization as potential treatment–preliminary study in pigs. *Radiology* 2008;**246**(3):783–9.
18. Carnevale FC, Antunes AA, da Motta Leal Filho JM, de Oliveira Cerri LM, Baroni RH, Marcelino AS, et al. Prostatic artery embolization as a primary treatment for benign prostatic hyperplasia: preliminary results in two patients. *Cardiovasc Interv Radiol* 2010;**33**(2):355–61.
19. Bilhim T, et al. Prostatic arterial supply: anatomic and imaging findings relevant for selective arterial embolization. *J Vasc Interv Radiol* 2012;**23**(11):1403–15.
20. Bilhim T, et al. Does polyvinyl alcohol particle size change the outcome of prostatic arterial embolization for benign prostatic hyperplasia? Results from a single-center randomized prospective study. *J Vasc Interv Radiol* 2013;**24**:1595–602.
21. Pereira JA, et al. Radiologic anatomy of arteriogenic erectile dysfunction: a systematized approach. *Acta Med Port* 2013;**26**(3):219–25.
22. Pisco JM, et al. Erratum to: embolisation of prostatic arteries as treatment of moderate to severe lower urinary symptoms (LUTS) secondary to benign hyperplasia: results of short- and mid-term follow-up. *Eur Radiol* 2013;**23**(9):2573–4.
23. Pisco JM, et al. Embolisation of prostatic arteries as treatment of moderate to severe lower urinary symptoms (LUTS) secondary to benign hyperplasia: results of short- and mid-term follow-up. *Eur Radiol* 2013;**23**(9):2561–72.

24. Pais D, et al. Sciatic nerve high division: two different anatomical variants. *Acta Med Port* 2013;**26**(3):208–11.
25. Bilhim T, et al. Middle rectal artery: myth or reality? Retrospective study with CT angiography and digital subtraction angiography. *Surg Radiol Anat* 2013;**35**(6):517–22.
26. Bilhim T, et al. Unilateral versus bilateral prostatic arterial embolization for lower urinary tract symptoms in patients with prostate enlargement. *Cardiovasc Interv Radiol* 2013;**36**(2):403–11.
27. Bilhim T, et al. Radiological anatomy of prostatic arteries. *Tech Vasc Interv Radiol* 2012;**15**(4):276–85.
28. Bouissou H, Talazac A. Arterial vascularization of the normal and the pathological prostate. *Ann Anat Pathol (Paris)* 1959;**4**(1):63–79.
29. Flocks RH. The arterial distribution within the prostate gland; its role in transurethral prostatic resection. *J Urol* 1937;**37**:524–48.
30. Ambrosio JD, De Almeida JS, De Souza A. Origin of prostatic arteries in man. *Rev Paul Med* 1980;**96**(3–4):52–5.
31. Bilhim T, et al. Prostatic arterial supply: demonstration by multirow detector angio CT and catheter angiography. *Eur Radiol* 2011;**21**(5):1119–26.
32. Pereira JA, et al. Patient selection and counseling before prostatic arterial embolization. *Tech Vasc Interv Radiol* 2012;**15**(4):270–5.
33. Carnevale FC, Antunes AA. Prostatic artery embolization for enlarged prostates due to benign prostatic hyperplasia. How I do artery it. *Cardiovasc Interv Radiol* 2013;**36**:1452–63.
34. Antunes AA, et al. Clinical, laboratorial, and urodynamic findings of prostatic artery embolization for the treatment of urinary retention related to benign prostatic hyperplasia. A prospective single-center pilot study. *Cardiovasc Interv Radiol* 2013;**36**(4):978–86.
35. Bienz M, Hueber PA, Al-Hathal N, McCormack M, Bhojani N, Trinh QD, et al. Accuracy of transrectal ultrasonography to evaluate pathologic prostate weight: correlation with various prostate size groups. *Urology* 2014;**84**(1):169–74.
36. Wang M, Guo L, Duan F, Yuan K, Zhang G, Li K, et al. Prostatic arterial embolization for the treatment of lower urinary tract symptoms as a result of large benign prostatic hyperplasia: a prospective single-center investigation. *Int J Urol* 2015;**22**(8):766–72.
37. Russo GI, Kurbatov D, Sansalone S, Lepetukhin A, Dubsky S, Sitkin I, et al. Prostatic arterial embolization vs open prostatectomy: a 1-year matched-pair analysis of functional outcomes and morbidities. *Urology* 2015;**86**(2):343–8.
38. Kurbatov D, Russo GI, Lepetukhin A, Dubsky S, Sitkin I, Morgia G, et al. Prostatic artery embolization for prostate volume greater than 80 cm^3: results from a single-center prospective study. *Urology* 2014;**84**(2):400–4.
39. Gao YA, Huang Y, Zhang R, Yang YD, Zhang Q, Hou M, et al. Benign prostatic hyperplasia: prostatic arterial embolization versus transurethral resection of the prostate–a prospective, randomized, and controlled clinical trial. *Radiology* 2014;**270**(3):920–8.
40. Li Q, Duan F, Wang MQ, Zhang GD, Yuan K. Prostatic arterial embolization with small sized particles for the treatment of lower urinary tract symptoms due to large benign prostatic hyperplasia: preliminary results. *Chin Med J (Engl)* 2015;**128**(15):2072–7.
41. Pisco J, et al. Prostatic arterial embolization for benign prostatic hyperplasia: short- and intermediate-term results. *Radiology* 2013;**266**(2):668–77.
42. Jones P, Rai BP, Nair R, Somani BK. Current status of prostate artery embolization for lower urinary tract symptoms: review of world literature. *Urology* 2015;**86**(4):676–81.
43. Feng S, Tian Y, Liu W, Li Z, Deng T, Li H, et al. Prostatic arterial embolization treating moderate-to-severe lower urinary tract symptoms related to benign prostate hyperplasia: a meta-analysis. *Cardiovasc Interv Radiol* 2017;**40**(1):22–32.

44. Tinto H, et al. Rio prostatic hyperplasia: short and medium prostatic artery follow-up. *Embolization Tech Vasc Interv Treatment Radiol Benign* 2012;**15**(4):290–3.
45. Jones P, Rai BP, Aboumarzouk OM, Somani BK. Prostatic urethral lift vs prostate arterial embolization: novel nonablative strategies in the management of lower urinary tract symptoms secondary to benign prostate hyperplasia. *Urology* 2016;**87**:11–7.
46. Fernandes L, et al. Prostatic arterial embolization: post-procedural follow-up. *Tech Vasc Interv Radiol* 2012;**15**(4):294–9.
47. Moreira AM, et al. Transient ischemic rectitis as a potential complication after prostatic artery embolization: case report and review of the literature. *Cardiovasc Interv Radiol* 2013;**36**(6):1690–4.

Further Reading

1. Valdivieso O'Donova R, Huber PA, Bhojani N, Trinh QD, Zorn K. Prostatic artery embolization. In: Chughtai B, Te AE, Kaplan SA, editors. *Treatment of benign prostatic hyperplasia: modern alternative to transurethral resection of the prostate.* 1st ed. New York: Springer; 2015.
2. Bagla Interv S, et al. Utility of cone-beam CT imaging in prostatic artery embolization. *J Vasc Radiol* 2013;**24**:1603–7.
3. McVary KT. BPH: epidemiology and comorbidities. *Am J Manag Care* 2006;**12**(5 Suppl.):S122–8.

Ethanol Injection of the Prostate in Benign Prostatic Hypertrophy

21

Douglas C. Cheung, Alexandra L. Millman, Matthew Lenardis, Dean Elterman
University of Toronto, Toronto, ON, Canada

Background

Benign prostatic hypertrophy (BPH) is a common cause of morbidity among aging men worldwide.[1–3] Complications of BPH include renal dysfunction, recurrent urinary tract infections, stones, hematuria, and urinary retention.[4,5] For the treatment of medically refractory BPH and/or BPH requiring surgical intervention due to complications, monopolar transurethral resection of prostate (TURP) continues to be the gold standard therapy.[6,7]

However, in recent years, alternative modalities have continued to gain further momentum as less invasive strategies to manage BPH.[7–10] Ideal candidates for these procedures include patients who are poor surgical candidates and those with numerous comorbidities. Since they may have difficulty tolerating a TURP, these patients would be best suited by a shorter procedure completed under local/regional anesthetic on an outpatient basis. In this regard, several injectable agents have been studied to determine their efficacy in the management of BPH. One such technology is ethanol ablation of the prostate.

Historical Application

Injecting substances into the prostate has long been described for the treatment of prostatitis, infection, and lower urinary tract symptoms (LUTS).[11,12] The earliest descriptions date back to the early 19th century when Sir B. C. Broadie described treatment of prostatic abscesses using a transperineal puncture. This approach was further expanded in 1896 when Picard stated that prostatic abscesses were further accessible via a transrectal or transurethral approach. These approaches to treatment of prostatic infection continued to persist through the turn of the century.

During this same period, intraprostatic injections were also being evaluated for the management of LUTS. In the early 1890s, Hardinge described using a mixture of carbolic acid, distilled water, glacial acetic acid, and glycerin injected into the prostate for management of obstructive urinary symptoms. This was later expanded by Talwar and Pande who in 1966 underwent the first large cohort description of intraprostatic injection using a similar combination of carbolic acid, glacial acetic acid, and glycerin to

A Comprehensive Guide to the Prostate. https://doi.org/10.1016/B978-0-12-811464-3.00021-0

treat 188 patients with LUTS. Many of these study participants were designated to be higher risk candidates. They reported favorable results as compared to the then-standard traditional retropubic prostatectomy, helping to popularize prostatic injections as a viable approach to management of these symptoms.[13]

Ethanol injection into the prostate as a form of ablative therapy was then first described in canines by Gray et al. in 1988.[14] The first techniques involved transperineal injection into the prostate but were initially marred by difficulties with sphincter necrosis and were felt to require ultrasound guidance. In 1999, Plante et al. also described transurethral injection of the prostate gland, again in a canine model.[15] They performed a number of superficial injections in a systematic circumferential manner within the prostatic urethra at a single point, resulting in significant focal necrosis, cavity formation, and a subsequently widened prostatic urethra. Encouragingly, there was no capsular penetration and the degree of systemic absorption was noted to be minimal throughout the study. In addition to these techniques, transrectal injection has now been described in a limited number of studies. Most recently in 2006, a small randomized controlled study examining the effect of transurethral injection of ethanol into the prostate demonstrated favorable results with no serious adverse events.[16]

Mechanism of Action

The mechanism of action of ethanol injection remains to be fully elucidated. The predominant theory is that ethanol causes fibrotic and coagulative necrosis through protein denaturation and cell membrane lysis.[17] The resultant effect is the destruction of the glandular epithelium and basement membrane, in turn producing local necrosis, cavitation of the prostate, and resulting enlargement of the prostatic urethra.[18] However, destruction of the afferent sensory nerves has also been postulated to reduce the symptomatic burden of BPH.

Transurethral Injection

Transurethral injection has been the most extensively studied method of delivery. Variable methods of injection have been described, varying according to prostatic volume, length, and median lobe.[19] Intraprostatic injection across studies varies significantly in volume (between 2 and 26 cc), using a curved or straight needle, distance along the prostatic urethra, and circumferential location within the urethra.[20] Most commonly, the ProstaJect device employs a retractile 22 gauge curved needle through a cystoscopic sheath to deliver the ethanol injection.[21] However, regardless of differences in injection location and volume, numerous studies have demonstrated initial promising results for intraprostatic ethanol injection.

Goya et al. first popularized their experience with transurethral injection in 1999—their results were durable to the medium term and after a follow-up of 3 years, patients continued to show reduced International Prostate Symptom Score (IPSS) and flow rate improvements.[22] 59% of patients did not require further treatment and no major complications were noted in the study. These early series demonstrated an initial strong safety profile with good tolerance and reasonable effectiveness for BPH therapy in a moribund population.

Despite promising short-term data, questions remained about its efficacy as a long-term treatment modality for BPH as 41% of cases above already required retreatment by the end of 3 years.[22] As such, subsequent studies continued to investigate the medium and long-term effects of transurethral intraprostatic ethanol injection. The longest durations of follow-up have been described by Buchholz et al. and Fayed et al. after 54 and 50 months of follow-up, respectively.[23,24] In the study by Buchholz et al. 56 men underwent transurethral intraprostatic injection.[23] They demonstrated a durable reduction in IPSS scores (17 vs. 10, preoperatively vs. postoperatively, respectively), maximal flow rate improvement (8 vs. 11 cc/s), decreased postvoid residuals (180 vs. 68 cc), and decreased prostatic volume (53 vs. 38 cc). Repeat intervention was required in 23% of patients and no mortality was noted. However, a major limitation of this study was the loss of patients to follow-up: only 14/56 (25%) of the total cohort completed the 4 year duration.

Equally, Fayed et al. described the 50 months of follow-up results of intraprostate ethanol injection with 25 of 35 (73%) patients completing follow-up to over 4 years.[24] Similar to the study by Buchholz et al. these patients had consistent improvements in IPSS scores (22 vs. 10, preoperatively vs. postoperatively, respectively), maximal flow rate (6 vs. 17 cc/s), postvoid residuals (69 vs. 36 cc), and prostatic volume (53 vs. 50 cc), which was related to the amount and number of sites of ethanol injection. Only 9/35 (26%) of patients required further intervention. However, 5 of 35 (13%) patients developed complications with stress incontinence (n = 1), infection (urinary tract infection, epididymo-orchitis, n = 1 each), urethral stricture (n = 1), and hematuria (n = 1). There was no mortality noted.

To date, the most recent study demonstrated 52 patients who underwent transurethral ethanol injection in 2014 in Turkey and reported improvement in symptoms scores and flow rates.[18] Minimal adverse events were noted including hematuria (not requiring transfusion) and infection (epididymo-orchitis). Other contemporary studies worldwide have also corroborated these results in regard to LUTS reduction, urinary flow, and prostatic size improvements.[19,25,26] These studies have reported complications of irritative LUTS, hematuria, urinary retention, and infection attributable to the ethanol injection in 1-year follow-up.[19,25,26]

Perhaps most significantly, Mirone et al. documented their experience in a European multicentre international study of 115 patients in 2004.[21] This continues to be the largest cohort to date for intraprostatic ablative ethanol injection and revealed substantial concerns regarding its safety and tolerability for use. Follow-up after 1 year revealed IPSS reduction (21 vs. 10, preoperatively vs. postoperatively, respectively), maximal flow rate increase (10 vs. 13 cc/s), prostatic size reduction (46 vs. 39 cc). Again, the most common adverse events were irritative LUTS (26%), urinary retention (17%), and hematuria (16%). Other side effects included urinary tract infection, pain, urinary incontinence, hematospermia, erectile dysfunction, anejaculation, meatal stenosis, and hypotension. Significantly, two patients experienced serious complications as a result of bladder necrosis following treatment requiring urinary diversion/cystectomy and ureteral reimplantation. Although this resulted in a procedural modification (felt to be secondary to proximal injection near the bladder neck), this remains a critical adverse effect to be aware of and casts doubt regarding the safety profile of intraprostatic ethanol injection.

Transperineal Injection

As an alternative to transurethral ethanol injection, a transperineal approach has been speculated as a method to increase tolerability and minimize the invasive nature of these injections. Belgrano et al. continued to note significantly improvements in maximal flow rates and IPSS scores in patients at 6 months of follow-up.[27] American Urological Association (AUA) Symptom Scores decreased (21 vs. 10, preoperatively vs. postoperatively, respectively) and maximal flow rate improved (11 vs. 16 cc/s) in patients with BPH treated with transperineal prostatic ethanol injection. Patients tolerated the procedure very well with described mild hematuria and urge incontinence, but no severe adverse events.

A subsequent study by Chiang et al. continued to demonstrate tolerability with only mild LUTS and one event of retention.[28] Again, the patients demonstrated significant improvement to their symptom scores and flow rates. Unfortunately, both studies were markedly limited by small sample sizes (8 patients in the Belgrano study and 11 patients in the Chiang study). Furthermore, concerns about backflow through the injection tract and necrosis of the external sphincter have plagued this modality of injection.

Transrectal Injection

Recently, further studies have also evaluated a transrectal approach of ethanol injection: in 2014, Dong et al. studied 70 patients followed over 2 years with transrectal injection of ethanol under transrectal ultrasound image guidance.[29] They found persistent improvements in IPSS scores (29 vs. 9, preoperatively vs. postoperatively, respectively) maximal flow rates (4.7 vs. 15.3 cc/s) and decreased prostatic volume. Adverse events were usually minor including dysuria/abdominal pain, infection, and requirement for catheterization. These results are similar to an earlier study by Larson et al. in 2006, with similar improvements amongst all groups including transrectal, transperineal, and transurethral injection.[30] No studies have noted urethral–rectal fistulas, other serious morbidities, or mortality to date.

Safety

As above, the suitability and safety profile of ethanol as an injectable substrate for prostatic ablation remains unclear. Administration through a transurethral, transperineal, and transrectal approach have all been studied with no definitive consensus on the delivery route, although all have been described under local and/or regional anesthesia. Encouragingly, this confers intraprostatic injection a potential suitability for patients not amenable to surgical therapy or more invasive techniques such as TURP. Accordingly, most of the published literature has appropriately included patients with significant comorbidities with no described mortality to date.

Unfortunately, however, major concerns remain regarding ethanol ablative therapy. Backflow through the needle tract continues to be an issue, leading to decreased delivery effectiveness and possible increased morbidity of the procedure. Additionally, the majority of the data reflect outcomes from short-term studies, including the most frequent adverse events of abdominal pain, LUTS including dysuria, infection, and hematuria.

Most worrying, however, is the accidental misinjection of the ethanol, particularly into the bladder (bladder necrosis) or external sphincter (incontinence).[31] In the largest multicentre international European study, 2 out of 115 patients required cystectomy/bladder diversion and ureteral reimplantation due to inadvertent proximal injection.[21]

Guideline Consensus

Currently, there remain no guideline recommendations for intraprostatic ethanol ablative therapy for the treatment of BPH. Most recently, the European Association of Urology published a revised guideline in 2015, which documented that intraprostatic ethanol injections remain "experimental and should only be performed in clinical trials" and was given a grade C recommendation (level 3 evidence).[20] The AUA did not include ethanol injection within their guideline statements, and it remains not recommended for standard therapy under the Canadian Urological Association guidelines.[7,32]

Further Direction

Most recently in the literature, different needle catheters for use have been evaluated with regard to better diffusion properties of ethanol injection in cadaveric prostates.[33] In another study, Larson et al. tackled the difficulty in applying a liquid-based alcohol with a gel-based injection to decrease run-off into undesired treatment areas.[30] However, these technologies have yet to catch on.

Additional studies are therefore necessary to evaluate the application methodology and clinical efficacy of intraprostatic ethanol injections before the recommendation of intraprostatic ethanol ablative therapy becomes standard treatment for BPH. At this point, the ideal technique of injection is unknown: existing studies have no clear consensus and have described a variety of approaches and techniques (locations and number of injections), depending on the prostatic volume, length, and median lobe presence.

References

1. Vuichoud C, Loughlin KR. Benign prostatic hyperplasia: epidemiology, economics and evaluation. *Can J Urol* 2015;(22 Suppl. 1):1–6.
2. Lepor H. Nonoperative management of benign prostatic hyperplasia. *J Urol* 1989;**141**(6): 1283–9.
3. Egan KB. The epidemiology of benign prostatic hyperplasia associated with lower urinary tract symptoms: prevalence and incident rates. *Urol Clin N Am* 2016;**43**(3):289–97.
4. Kim EH, Larson JA, Andriole GL. Management of benign prostatic hyperplasia. *Annu Rev Med* 2016;**67**:137–51.
5. Hollingsworth JM, Wilt TJ. Lower urinary tract symptoms in men. *BMJ* 2014;**349**:g4474.
6. Taylor BL, Jaffe WI. Electrosurgical transurethral resection of the prostate and transurethral incision of the prostate (monopolar techniques). *Can J Urol* 2015;(22 Suppl. 1):24–9.
7. McVary KT, et al. Update on AUA guideline on the management of benign prostatic hyperplasia. *J Urol* 2011;**185**(5):1793–803.
8. Perera M, et al. Prostatic urethral lift improves urinary symptoms and flow while preserving sexual function for men with benign prostatic hyperplasia: a systematic review and meta-analysis. *Eur Urol* 2015;**67**(4):704–13.
9. Nair SM, Pimentel MA, Gilling PJ. Evolving and investigational therapies for benign prostatic hyperplasia. *Can J Urol* 2015;(22 Suppl. 1):82–7.
10. Lusuardi L, Hruby S, Janetschek G. New emerging technologies in benign prostatic hyperplasia. *Curr Opin Urol* 2013;**23**(1):25–9.
11. Saemi AM, Folsom JB, Plante MK. Injection therapy for prostatic disease: a renaissance concept. *Indian J Urol* 2008;**24**(3):329–35.
12. Plante MK, Folsom JB, Zvara P. Prostatic tissue ablation by injection: a literature review. *J Urol* 2004;**172**(1):20–6.
13. Talwar GL, Pande SK. Injection treatment of enlarged prostate. *Br J Surg* 1966;**53**(5):421–7.
14. Littrup PJ, et al. Percutaneous ablation of canine prostate using transrectal ultrasound guidance. Absolute ethanol and Nd:YAG laser. *Invest Radiol* 1988;**23**(10):734–9.
15. Zvara P, et al. Ablation of canine prostate using transurethral intraprostatic absolute ethanol injection. *Urology* 1999;**54**(3):411–5.
16. Plante MK, et al. Phase I/II examination of transurethral ethanol ablation of the prostate for the treatment of symptomatic benign prostatic hyperplasia. *J Urol* 2007;**177**(3):1030–5. discussion 1035.
17. Andersson KE. Treatment of lower urinary tract symptoms: agents for intraprostatic injection. *Scand J Urol* 2013;**47**(2):83–90.
18. Arslan M, et al. Primary results of transurethral prostate ethanol injection. *Int Urol Nephrol* 2014;**46**(9):1709–13.
19. Magno C, et al. Transurethral ethanol ablation of the prostate (TEAP): an effective minimally invasive treatment alternative to traditional surgery for symptomatic benign prostatic hyperplasia (BPH) in high-risk comorbidity patients. *Int Urol Nephrol* 2008;**40**(4):941–6.
20. Gratzke C, et al. EAU guidelines on the assessment of non-neurogenic male lower urinary tract symptoms including benign prostatic obstruction. *Eur Urol* 2015;**67**(6):1099–109.
21. Grise P, et al. Evaluation of the transurethral ethanol ablation of the prostate (TEAP) for symptomatic benign prostatic hyperplasia (BPH): a European multi-center evaluation. *Eur Urol* 2004;**46**(4):496–501. discussion 501-2.
22. Goya N, et al. Transurethral ethanol injection therapy for prostatic hyperplasia: 3-year results. *J Urol* 2004;**172**(3):1017–20.

23. El-Husseiny T, Buchholz N. Transurethral ethanol ablation of the prostate for symptomatic benign prostatic hyperplasia: long-term follow-up. *J Endourol* 2011;**25**(3):477–80.
24. Sakr M, et al. Transurethral ethanol injection therapy of benign prostatic hyperplasia: four-year follow-up. *Int J Urol* 2009;**16**(2):196–201.
25. Faruque MS, et al. Evaluation of transurethral ethanol ablation of prostate for symptomatic benign prostatic hyperplasia. *Mymensingh Med J* 2012;**21**(2):265–9.
26. Mutaguchi K, et al. Transurethral ethanol injection for prostatic obstruction: an excellent treatment strategy for persistent urinary retention. *Urology* 2006;**68**(2):307–11.
27. Savoca G, et al. Percutaneous ethanol injection of the prostate as minimally invasive treatment for benign prostatic hyperplasia: preliminary report. *Eur Urol* 2001;**40**(5):504–8.
28. Chiang P, et al. Pilot study of transperineal injection of dehydrated ethanol in the treatment of prostatic obstruction. *Urology* 2003;**61**(4):797–801.
29. Li Y, Zhao Q, Dong L. Efficacy and safety of ultrasound-guided transrectal ethanol injection for the treatment of benign prostatic hyperplasia in patients with high-risk comorbidities: a long-term study at a single tertiary care institution. *Urology* 2014;**83**(3):586–91.
30. Larson BT, et al. Intraprostatic injection of alcohol gel for the treatment of benign prostatic hyperplasia: preliminary clinical results. *ScientificWorldJournal* 2006;**6**:2474–80.
31. Chung A, Woo HH. What's truly minimally invasive in benign prostatic hyperplasia surgery? *Curr Opin Urol* 2014;**24**(1):36–41.
32. Nickel JC, et al. 2010 update: guidelines for the management of benign prostatic hyperplasia. *Can Urol Assoc J* 2010;**4**(5):310–6.
33. King BJ, et al. Intraprostatic ethanol diffusion: comparison of two injection methods using ex vivo human prostates. *Prostate Cancer Prostatic Dis* 2015;**18**(3):237–41.

Future Surgical Procedures: iTind, Rezūm, and Aquablation

22

James P. Winebrake, Dominique Thomas, Alexis Te, Bilal Chughtai
Weill Cornell Medicine, New York, NY, United States

Introduction

Benign prostatic hyperplasia (BPH) presenting with lower urinary tract symptoms (LUTS) is a source of significant morbidity and quality-of-life (QoL) reduction worldwide,[1] investigating cost-effective strategies for treatment of BPH/LUTS with minimization of adverse effects is an important health concern. Furthermore, clinical ineffectiveness, lifetime treatment cost, and adverse effects of combination pharmacological therapies—the latter including erectile and ejaculatory dysfunction, postural hypotension, peripheral edema, dyspnea, somnolence, etc...[2]—contribute to the estimated quarter or more of BPH/LUTS patients who voluntarily discontinue medical therapy.[3,4] Alternatively, surgical intervention, though effective for LUTS, is associated with a great deal of surgical and perioperative morbidity.[5] Likewise, transurethral resection of prostate (TURP), which has been regarded as the gold standard, risks complications such as perioperative bleeding, prolonged catheterization, urinary incontinence, retrograde ejaculation, etc.[6] It follows that the development of novel minimally invasive techniques (MITs) these therapies generally have more acceptable complication profiles, though their efficacy is less well established.[7] The MITs category comprises transurethral incision, thermoablative techniques, mechanical maneuvers, and a variety of other novel techniques.[7]

The aim of this chapter will be to introduce three investigative MITs: heat- and resection-free implantation of Temporary Implantable Nitinol Device (iTind) (MediTate), convective thermal vapor energy application by Rezūm System (NxThera, Inc.), and high-pressure water-jet ablation under robotic image-guided control via the AquaBeam System (PROCEPT BioRobotics Corp.).

iTind: Technique and Clinical Results

The iTind implant is crafted from nitinol alloy and measures 50 mm by 33 mm at maximum length and diameter. Its main purpose is to restore urethral function by temporary implantation without heat application or tissue removal. With the patient under light sedation, the bladder is filled with saline solution and the device is inserted in a folded configuration via urethrocystoscopy. An anchoring leaflet is oriented at the 6 o'clock position on the bladder neck, securing the device for the duration of

A Comprehensive Guide to the Prostate. https://doi.org/10.1016/B978-0-12-811464-3.00022-2

implantation; the device is retracted distally from the bladder to form sites of contact between longitudinal struts and obstructive prostatic tissue. Withdrawal of the hollow delivery apparatus allows the device to passively expand further, in doing so providing a mechanism for minimally invasive dilation of the bladder neck and prostatic urethra. This is accomplished via exertion of radial force by the struts at the 12, 5, and 7 o'clock positions. Ischemic necrosis and incision occur over the course of the 5-day implantation interval, thereby reshaping tissue to provide channels for urinary flow and relief of BPH symptoms. The iTind is removed by Foley catheter or cystoscope in the outpatient setting with the use of topical anesthesia.[8]

Porpiglia et al. conducted the first reported clinical experience with iTind for BPH-associated LUTS. Thirty-two patients with LUTS were enrolled in a single arm, prospective study to assess device feasibility and safety. Inclusion criteria was patient age (>50 years), presentation of BPH symptoms (International Prostate Symptom Score [IPSS] ≥10, Q_{max} ≤12 mL/s), and prostate size by transrectal ultrasonography (TRUS) (<60 mL). Postimplantation pain scores were recorded using a visual analog scale (VAS). Patients' status was assessed by uroflowmetry, IPSS, and IPSS QoL index at baseline, 5 days (the device removal visit), 3 weeks, 6 weeks, 3 months, 6 months, and 12 months following implantation. Patients were asked by "Overall, how satisfied are you with treatment you received for your prostate disease intervention?" responsed on a 5-point Likert scale for satisfaction. Early and late post-op adverse events were recorded.

Postoperative stay was a mean (SD) operative time of 2 (1) min and median (IQR) length of stay of 1 (1–2) days. The latter result was perhaps artificially inflated due to a conservative approach in the trial's early stages; the final 12 patients were discharged on the day of the surgery. Median (IQR) VAS score and acetaminophen use at 6 h postoperatively was 2 (2–4) and 1000 mg, respectively. These outcomes, as well as the lack of postoperative catheterization and unplanned readmission during the 5-day implantation period, demonstrated good device tolerance. Though no late complications (greater than 30 days) were noted, four patients (12.5%) experienced urinary incontinence at day 1 following improper placement and dislodging of the TIND; urinary retention at day 0; urinary tract infection (UTI) at 2 weeks; 2-cm prostatic abscess at 4 weeks atrial fibrillation, and hyperglycemia in a patient with type II diabetes.[8]

Furthermore, statistically significant improvements in Q_{max} and IPSS with respect to baseline were recorded within 3 weeks of device removal. By the 6-week mark mean improvement to Q_{max} was over 60%; this result was maintained at each subsequent follow-up visit. The mean IPSS percent decreases from baseline peaked at approximately 55% (3 months) and remained over 40% at 1 year. QoL scores improved similarly and significantly over the course of the trial. Medical therapy was discontinued for all enrolled patients at 3 months postoperatively. At 12 months, 26 patients (82%) were "extremely satisfied" or "satisfied" with the procedure, while 5 were "uncertain" and 1 was "dissatisfied," yielding median (IQR) 5-point Likert scale scores of 5 (4–5). There was also no need for medical treatment or surgical intervention during the trial period.[8]

Porpiglia et al. found iTind to be safe and feasible when executed properly, demonstrating clinical results rivaling or exceeding those of other novel minimally invasive therapies larger studies with placebo are currently underway.

Rezūm: Technique and Clinical Results

The Rezūm System (NxThera, Inc.) is designed to provide thermoablative therapy to the hyperplastic prostate via electromagnetic induction of radiofrequency power. However, unlike TUNA or other techniques in this category, Rezūm delivers power generated in the handle of the transurethral device in a sterile water vapor medium at 103°C (Convective Water Vapor Energy: WAVE).[9] In comparison to microwave, radiofrequency, and interstitial laser conductive thermal techniques, this novel approach exposes tissues to similar quantities of heat for much shorter durations of time. Although Rezum is not limited by a middle lobe, each zone must be individually treated since its therapy cannot penetrate tissue barriers between zones.

After the patient is sedated, direct visualization of the urethra is obtained by cystoscopy. A polyether ether ketone needle is inserted into the transition zone of the prostate and 0.5 mL of vapor is injected circumferentially from the needle tip.[9] A single 9-s injection produces a spherical lesion ranging from approximately 1.5–2.0 cm^3 in size.[3] Vapor disperses throughout interstitial space by convection and condenses to liquid form, transmitting up to 208 cal of thermal energy per injection directly to cell membrane surfaces (540 calories per gram of vapor). Dispersion of energy allows for delivery that is more efficient than by conductive means,[10] and elimination of a thermal gradient that risks tissue charring and desiccation.[11] However, that heat transfer remains sufficient to induce immediate cell death occurring at temperatures ranging between 70 and 80°C. The needle is retracted and repositioned approximately 1 cm distally along the urethra for each subsequent injection, forming adjoining pockets of ablated tissue as needed based on preoperative evaluation.[9]

One of the first trials evaluating efficacy of Rezūm's convective vapor energy transfer system in humans, included patients with BPH/LUTS who underwent adenectomy immediately following the procedure. Following excision of the prostate, metabolically active tissue alongside well-circumscribed areas of necrosis was seen on gross exam.[9]

Soon afterward in a multicenter study evaluating physical changes on imaging following Rezūm, Mynderse et al. followed 44 patients recently treated with the therapy using serial gadolinium-enhanced magnetic resonance imaging. Ablative lesions were surveilled at baseline (1 week postoperatively) for up to 6 months. Lesions were nearly resolved at the 3- and 6-month follow-up; mean volume decreased by 91.5%, then 95.1%. Furthermore, prostate size was reduced by 28.9% and transition zone volume diminished by 38.0%. Importantly, operative lesions were contained in the transition zone without penetrating tissue boundaries or affecting other nearby organ structures.[12]

Roehrborn et al. conducted a 2-year prospective, 15-center, double-blind randomized controlled trial of nearly 200 men with moderate to severe symptomatic BPH. Randomization to Rezūm and control arms was stratified by site in a 2:1 ratio. All patients were assessed at baseline and follow-up by IPSS, QoL, incontinence, and sexual performance measures. Whereas the Rezūm arm patients were monitored over a period of 2 years, controls were unblinded 3 months after performance of a sham procedure before undergoing Rezūm therapy and tracking for approximately 1 additional year in a crossover study. By blinding patient and follow-up provider following sham procedure and until crossover, the study format offered opportunity for paired self-control comparison.[3] This was a notable strength given placebo effect associated with medical and surgical interventions for BPH.[13]

Patients who initially underwent Rezūm therapy experienced sustained improvement at 6, 12, and 24 months postoperatively. Mean IPSS was reduced by at least 51% ($P<.0001$) with 74% experiencing IPSS decreases of 8 points or greater at 24 months. Mean Q_{max} and QoL improvements remained at least 50% over the 24-month course ($P<.0001$) without significant decreases in erectile or ejaculatory function,[3,14] also noted urinary storage and voiding functions were significantly improved at 1 month and stable throughout the trial period ($P<.0001$). Furthermore, patients in the control arm experienced significant reductions in IPSS from time of crossover to postoperative assessments ($P=.0004$) at rates similar to those treated initially. QoL and Q_{max} also improved significantly in the treatment group.[3]

As an aside, perioperative pain management strategies in the literature varied between and within cohorts. In the above trial oral sedation alone was sufficient for 69% of patients, while 10% received conscious sedation and 21% received a prostate block. Three serious adverse events were recorded in two crossover patients: bladder neck contracture and bladder calculi (6-month follow-up) and urosepsis after follow-up cystoscopy. Various expected nonserious adverse events occurred in crossover and noncrossover patients, including dysuria, hematuria, urinary retention, suspected UTI, etc. at relatively low rates.[3]

Mild-to-moderate adverse effects were generally limited.The efficacy and safety of this therapy make it "appropriate and low-risk treatment in the continuum between medical management and more invasive surgical approaches."[11] NxThera's Rezūm System received 510(k) clearance by the US FDA for treatment of BPH in August 2015.[15] Regardless, larger trials with prolonged follow-up will undoubtedly be needed in the determining the role of Rezūm in the BPH armamentarium.

Aquablation: Technique and Clinical Results

The PROCEPT AquaBeam System fills yet another void in the area of minimally invasive treatment for BPH with LUTS, notably by robotic, image-guided control without heat application. The system utilizes high-pressure water-jet technologies akin to those previously used in human liver and kidney resection, comprising a robotic handpiece with handle and a single-use AquaBlade cutting probe designed for transurethral insertion via rigid cystoscope. An inflatable balloon allows for anchoring distal to the bladder, while a fiberscope is integrated for direct visualization. The probe tip is precisely manipulated by axial rotation and longitudinal insertion/retraction to emit high-pressure saline jets orthogonally for selective glandular ablation that preserves vessels and the prostatic capsule. The last major component of the AquaBeam System is the external console that displays transverse and sagittal prostate scans obtained with a biplane TRUS probe in real time. The monitor and accompanying software also permits programmed customization of specifications for ablation based on existing anatomy and the extent of hyperplasia: length, sweep angle (up to 225°degrees), contour, and depth (up to 25 mm). Following tissue dissection, the console controls a lower pressure saline flow designed for guidance of a 2 W surgical laser for surface hemostasis.[16]

The first survival study of the Aquablation technique, completed by Faber et al. in canine models, yielded sufficiently patent lumen lined with intact, healthy epithelium on histology, as well as maintenance of capsule and other underlying structures. The procedure reduced operative time (Aquablation duration averaged 60.5 s; range 40–84 s), avoidance of thermal energy application, and finely tuned, computerized control. Limitations and complications including infection were attributed to the study's small cohort and dissimilarities between dog and human anatomy.[16]

In the following year, Gilling et al. published a first-in-man, single-center study of safety and feasibility in 15 patients with BPH/LUTS refractory to medical therapy. All patients reported a preoperative IPSS of >12, $Q_{max} \leq 12$ mL/s, and prostate size of 25–80 mL. All procedures were performed under general anesthesia and found to have succeeded from a technical standpoint, lacking serious or unexpected complication including blood loss and electrolyte imbalance. However, 8 of 15 patients experienced at least one of the following mild AEs common in MIT procedures within the 30-day postoperative window (Clavien-Dindo grade 1–11): dysuria (3/15), hematuria (3/15), pelvic discomfort (3/15), need for recatheterization (5/15), postoperative cardiac arrhythmia (1/15), and bladder spasms (1/15). No urinary incontinence, erectile dysfunction, or retrograde ejaculation were reported in any case based on International Index of Erectile Function and Incontinence Severity Index questionnaires. One patient did require a second procedure within 90 days dueinadequate resection.[17]

At the study's 6-month follow-up, mean IPSS score improved to 8.6 from a baseline of 23.1 ($P<.001$), while Q_{max} improved to 18.6 mL/s from 8.6 ($P<.001$). Mean detrusor pressure at Q_{max} was also measured, averaging 66 cmH2O at baseline and decreasing to 45 cmH2O at follow-up ($P<.05$). Mean prostate size assessed by TRUS was reduced by 31%–36 mL ($P<.001$). Moreover, researchers noted improvements to procedural efficiency with the following modifications in the final eight cases: use of a rollerball or loop for cautery as opposed to the previously described laser system (found to be relatively ineffective in prior cases, which utilized supplemental diathermy); addition of an aspiration pump for use during ablation; integration of live TRUS images on the system console for additional guidance. These adaptations reduced mean procedure time from 56 to 40 min.[17] MacRae and Gilling later wrote that a Foley catheter balloon tamponade of the prostatic cavity was ultimately preferred to produce effective hemostasis.[18]

Ploussard and Tollon commented on the first-in-man study's high temporary recatheterization rate and the aforementioned second required procedure in a single patient, reiterating the need for a larger sample and long-term follow-up.[19] Nevertheless, Cadeddu noted "impressive" functional results in the first-in-man study and the possibility that automated prostate ablation technology could "significantly alter clinical practice."[20]

The results of a larger multicenter trial of 57 patients, again executed by Gilling and colleagues, mirrored those above. Similar inclusion criteria were maintained, though patients with prostate volume of up to 100 mL were treated. However, clinical and safety assessment was continued up to 1 year following surgery (in 33 subjects as of the most recent reporting). All procedures were technically successful without major complication and mean operative and resection times were 38 and 7 min, respectively. Mild perioperative adverse events were temporary and occurred at rates similar to those reported for more traditional therapies for BPH. Again, no cases of retrograde ejaculation, urinary incontinence, or erectile dysfunction were recorded. In a comparison between values at baseline and most recent follow-up, IPSS decreased from 22.9 to 6.8, QoL from 5.0 to 1.6, and postvoid residual volume from 105 to 57 mL; Q_{max} improved from 7.8 to 16.7 mL/s at 12 months. Investigators found Aquablation to be a feasible, safe, and increasingly efficient modality of minimally invasive surgical intervention for LUTS associated with BPH, remarking that further study in randomized controlled trials was warranted.[21]

Despite encouraging results in inaugural preclinical and clinical studies, AquaBeam is EU EC Marked but currently limited to investigational use by the United States law. A multicenter, prospective randomized investigational device exemption clinical trial of 184 patients with comparison to TURP is ongoing as of current writing.[22] Results of this and additional trials will in large part dictate the value of Aquablation as a legitimate treatment option for BPH/LUTS patients.

Conclusion

The literature describes multitude therapies for BPH, both pharmacological and surgical. An increasing proportion of the latter are minimally invasive in nature; their development and potential integration alongside other strategies of the modern-day clinician is dependent on the results of ongoing research but likely worthwhile given the significant burden of disease on society. Three cutting-edge modalities stand out in particular and merit further study in larger-scale randomized controlled trials with lengthier follow-up: implantation of and urethral dilation by iTind, convective thermal vapor energy ablation by Rezūm, and robotically controlled water-jet ablation by AquaBeam. Each is characterized by a distinct profile of advantages and limitations, and importantly, the potential for emergence as a preeminent minimally invasive intervention for patients suffering from BPH/LUTS.

References

1. Parsons JK. Benign prostatic hyperplasia and male lower urinary tract symptoms: epidemiology and risk factors. *Curr Bladder Dysfunct Rep* 2010;**5**(4):212–8.
2. McConnell JD, Roehrborn CG, Bautista OM, et al. The long-term effect of doxazosin, finasteride, and combination therapy on the clinical progression of benign prostatic hyperplasia. *N Engl J Med* 2003;**349**(25):2387–98.
3. Roehrborn CG, Gange SN, Gittelman MC, et al. Convective thermal therapy: durable 2-year results of randomized controlled and prospective crossover studies for treatment of lower urinary tract symptoms due to benign prostatic hyperplasia. *J Urol* 2017;**197**(6):1507–16.

4. Verhamme KM, Dieleman JP, Bleumink GS, Bosch JL, Stricker BH, Sturkenboom MC. Treatment strategies, patterns of drug use and treatment discontinuation in men with LUTS suggestive of benign prostatic hyperplasia: the Triumph project. *Eur Urol* 2003;**44**(5):539–45.
5. Ferretti M, Phillips J. Prostatectomy for benign prostate disease: open, laparoscopic and robotic techniques. *Can J Urol* 2015;**22**(Suppl. 1):60–6.
6. Rocco B, Albo G, Ferreira RC, et al. Recent advances in the surgical treatment of benign prostatic hyperplasia. *Ther Adv Urol* 2011;**3**(6):263–72.
7. Christidis D, McGrath S, Perera M, Manning T, Bolton D, Lawrentschuk N. Minimally invasive surgical therapies for benign prostatic hypertrophy: the rise in minimally invasive surgical therapies. *Prostate Int* 2017;**5**(2):41–6.
8. Porpiglia F, Fiori C, Bertolo R, Garrou D, Cattaneo G, Amparore D. Temporary implantable nitinol device (TIND): a novel, minimally invasive treatment for relief of lower urinary tract symptoms (LUTS) related to benign prostatic hyperplasia (BPH): feasibility, safety and functional results at 1 year of follow-up. *BJU Int* 2015;**116**(2): 278–87.
9. Dixon CM, Rijo Cedano E, Mynderse LA, Larson TR. Transurethral convective water vapor as a treatment for lower urinary tract symptomatology due to benign prostatic hyperplasia using the Rezum((R)) system: evaluation of acute ablative capabilities in the human prostate. *Res Rep Urol* 2015;**7**:13–8.
10. Dixon CM, Cedano ER, Pacik D, et al. Two-year results after convective radiofrequency water vapor thermal therapy of symptomatic benign prostatic hyperplasia. *Res Rep Urol* 2016;**8**:207–16.
11. Woo HH, Gonzalez RR. Perspective on the Rezum(R) System: a minimally invasive treatment strategy for benign prostatic hyperplasia using convective radiofrequency water vapor thermal therapy. *Med Devices (Auckl NZ)* 2017;**10**:71–80.
12. Mynderse LA, Hanson D, Robb RA, et al. Rezum system water vapor treatment for lower urinary tract symptoms/benign prostatic hyperplasia: validation of convective thermal energy transfer and characterization with magnetic resonance imaging and 3-dimensional renderings. *Urology* 2015;**86**(1):122–7.
13. Welliver C, Kottwitz M, Feustel P, McVary K. Clinically and statistically significant changes seen in sham surgery arms of randomized, controlled benign prostatic hyperplasia surgery trials. *J Urol* 2015;**194**(6):1682–7.
14. McVary KT, Gange SN, Gittelman MC, et al. Erectile and ejaculatory function preserved with convective water vapor energy treatment of lower urinary tract symptoms secondary to benign prostatic hyperplasia: randomized controlled study. *J Sex Med* 2016;**13**(6):924–33.
15. 510(k) premarket notification: Rezum system. 2015. https://www.accessdata.fda.gov/scripts/cdrh/cfdocs/cfpmn/pmn.cfm?ID=K150786.
16. Faber K, de Abreu AL, Ramos P, et al. Image-guided robot-assisted prostate ablation using water jet-hydrodissection: initial study of a novel technology for benign prostatic hyperplasia. *J Endourol* 2015;**29**(1):63–9.
17. Gilling P, Reuther R, Kahokehr A, Fraundorfer M. Aquablation – image-guided robot-assisted waterjet ablation of the prostate: initial clinical experience. *BJU Int* 2016;**117**(6):923–9.
18. MacRae C, Gilling P. How I do it: aquablation of the prostate using the AquaBeam system. *Can J Urol* 2016;**23**(6):8590–3.
19. Ploussard G, Tollon C. Re: aquablation – image guided robotically-assisted waterjet ablation of the prostate: initial clinical experience. *Eur Urol* 2016;**69**(6):1157–8.

20. Cadeddu JA. Re: aquablation-image-guided robot-assisted waterjet ablation of the prostate: initial clinical experience. *J Urol* 2017;**197**(1):222.
21. Gilling P, Desai M, Anderson P, Tan A, Desai M. Mid-term results using aquablation, an image guided robot-assisted water jet ablation of the prostate, for the treatment of benign prostatic hyperplasia (BPH). *J of Urol* 2016;**195**(4S Suppl.):e458.
22. WATER Study. 2017. http://procept-biorobotics.com/water-study.php.

Further Reading

1. Lourenco T, Shaw M, Fraser C, MacLennan G, N'Dow J, Pickard R. The clinical effectiveness of transurethral incision of the prostate: a systematic review of randomised controlled trials. *World J Urol* 2010;**28**(1):23–32.

Index

Note: 'Page numbers followed by "f" indicate figures and "t" indicate tables'.

FSC
www.fsc.org
MIX
Papier | Fördert gute Waldnutzung
FSC® C083411

Kolibri 360 GmbH
Ferdinand-Jühlke-Straße 7
99095 Erfurt, Deutschland
produktsicherheit@kolibri360.de